Dunbar

Motorcycle
Collectibles

Schiffer
Publishing Ltd

77 Lower Valley Road, Atglen, PA 19310

Printed in China
ISBN: 0-88740-947-4

Book Design by Audrey L. Whiteside

Library of Congress Cataloging-in-Publication Data

Dunbar, Leila
    Motorcycle collectibles/Leila Dunbar.
       p.   cm.
    Includes bibliographical references and index.
    ISBN 0-88740-947-4 (pbk.)
    1. Motorcycles--Collectibles--Catalogs. 2. Motorcycling--United States--Collectibles--Catalogs. I. Title.
    TL440.D86   1996
    629.227'5'075--dc20   96-644
                            CIP

Published by Schiffer Publishing, Ltd.
77 Lower Valley Road
Atglen, PA 19310
Please write for a free catalog.
This book may be purchased from the publisher.
Please include $2.95 postage.
Try your bookstore first.

We are interested in hearing from authors
with book ideas on related subjects.

TABLE OF CONTENTS

ACKNOWLEDGMENTS
THANK YOU TO...

I would like to thank the following people for their contributions to this book: Doug, Debbie and Engla Leikala, for their generosity in allowing me into their house for a weekend to shoot photos of their tremendous collection and to discuss the merits of Rusty Wallace. I feel fortunate to have made three new friends so easily; John Cerelli, for helping to provide reference material and general information when needed; Ken Kalustian, for his photos of the Elvis Enthusiast Cover and Harley Neon Sign; Pat Simmons and Cris Sommer Simmons, for photos of their great collection of early sales literature and books and their enthusiastic introduction; Arline & Barry MacNeil, for helping round out a number of categories with their eclectic collection; Bob "Sprocket" Eckardt, who provided some very nice and rare pieces for the book; Randy Inman and James Julia Auctions, Inc., for photos and information; Howard Dunbar, who for once listened to his mother forty years ago and gave up motorcycling, because if he had kept going, neither he (nor I) might be here right now. Instead he started collecting cast iron toy cycles, which are shown in Chapter 11 of this book; Martha Dunbar, for her great meals and encouragement, both very nourishing to the spirit. Also, a special thanks to Sally Cort, personal editor and good friend for the past thirteen years, who has read and corrected all of my work with a warm pencil, an easy frankness, and an ever-present smile.

PREFACE
COLLECTING MOTORCYCLE MEMORABILIA

– PAT & CRIS SIMMONS

Time passes so quickly and the things that we have saved to remind us of the places we've been and the good times we've had begin to take on a whole new significance as the years roll on. As avid motorcycle enthusiasts, we both love and treasure all the various memorabilia that has been generated by this colorful and active sport.

To us, collecting is not just about pieces of paper and metal or about being materialists...to us, collecting is a passion... an honest-to-goodness passion for keeping in touch with our past. These old mementos have become our links to an era when life itself was so much simpler.

In the beginning, it was the motorcycles themselves that captured our imaginations — the wind in our faces, the sun at our backs, and the unbroken vista of hills and sky above a winding highway fading into the distance. Aboard our iron horses we felt the freedom the early cowboys must have experienced as they looked over the next rise at the beauty of the plains unfolding beyond. As our knowledge and enthusiasm grew, so did our interest in other aspects of our motorcycling hobby.

We have both been collectors of stuff since we were small children (everything from gum wrappers to postage stamps), but nothing has quite taken over our lives the way motorcycles have. From every wall of our house, from every cabinet or shelf, some image, sculpture, toy, tool, plate, cup, can, or piece of motorcycle advertising is visible. In every drawer and closet can be found an overabundance of t-shirts, vests, jackets, pants, chaps, and boots, all bearing some Indian or Harley logo or pin. In our file cabinets, *Harley Enthusiasts* and catalogues accumulate year after year, next to obscure advertising and tech manuals for long gone brands of forgotten two-wheeled mounts. The thought of throwing out something that someday might be a treasure to our kids or grandchildren is just too much to bear!

Looking through some of the old motorcycle magazines and brochures lets us relive the birth of the sport itself. These old pieces of paper allow us a peek into the past, way back as far as the early 1900s when motorcycling was new and exciting.

The sales brochures of that time depict men and women dressed in fancy riding clothes, cruising through life on a new, romantic journey, which of course included a new motorcycle! The color artwork of some of these sales brochures is unbelievably beautiful, as collectible for its historical significance as the original artwork itself. Looking back, you can actually trace the technology of the motorcycle engine throughout the years. Many improvements were added in those early years, things like electric lights. By the 1920s, motorcycles even came with a new feature called a front brake!

Much of the history of our country can also be traced through some of these old publications. Articles appearing in magazines such as *Colliers* in the early 1940s related what an important role the motorcycle played in the second world war. Year after year the world around us changed and the motorcycle was right there, changing and growing right along with the rest of the world.

This is, after all, the real reason we collect; not only for our love of the past and a longing for a time when life was simpler, but also for our kinship with an earlier era which we seek to preserve for others to enjoy.

We are just caretakers for the future generations of motorcycle enthusiasts who someday will carry on with their own particular approach to collecting.

Happy hunting....

Pat Simmons & Cris Sommer Simmons

Author's Note: I'm sure if they could, Cris Sommer Simmons and Pat Simmons would spend most of their time tooling about the country on their Harleys. However, Pat's full time gig as Doobie Brother guitarist and Cris's career as worldwide journalist and advocate for women's motorcycling, as well as keeping up with their three children, Lindsay, Josh, and Patrick, make huge demands on their time. Cris, the founder of *Harley Women Magazine,* also writes for *Hotbike Japan* and *Motorcycle Collector Magazine.* In 1994, she wrote a children's book about motorcycling, *Patrick Wants To Ride,* published by Steel Pony Press. A sequel, *Patrick's New Motorcycle Adventure,* will be out sometime in 1996. Also, Cris is one of four women profiled for a documentary on women's motorcycling that will air in 1996 on the Turner Broadcasting System.
Pat and Cris's longtime motomobilia collection includes early sales literature and books, bikes and toys. The lifelong passion that Cris and Pat share in their pursuit of motorcycling and collecting is an example of how enriching the sport/hobby of motorcycling can be.

INTRODUCTION

Out of the shadows, breaking the silence
Rumbling staccato exhaust, gleaming Harley
Lone rider, full leather
"The Moonlight Kid."

Riding the back roads of America
Embracing fraternity, breathing freedom
A smile, a wave, a swirl of dust
"The Moonlight Kid."

Poem by Howard & Leila Dunbar, copyright © 1993, all rights reserved

Who is the Moonlight Kid? It is the alter ego for anyone who has donned a leather jacket, hopped on a motorcycle and headed down the open road. The Moonlight Kid can be a bit wild but innocent, daring but shy. The Moonlight Kid loves people, but wants our founding fathers' tenets of freedom and independence. The Kid wants to live life on his or her own terms, finding the odd adventure that comes with travel, but never overstaying one's welcome. It's that part of us we would like to keep a bit mysterious.

It is odd that it is a closely knit pack of hog lovers who tightly grip the tenets of living free and at peace with oneself as an individual. It is a group that maintains its identity, which is found in its individuality.

It's even odder that the love for motorcycling crosses every class and strata of life. Who are the riders under the helmets that kick over and take off? How about the stereotypical guys that look like they could bench press their Harleys with their bellies and beards. Or, one can look for the bejeweled saddle bags of the rubies (rich urban bikers) — lawyers, doctors and businessman who shed suit coats for leather jackets for their weekend rebellions. Then, there's the rich legacy of celebrities who have helped make Milwaukee famous, from Clark Gable to Elvis to ZZ Top to Evel Knievel, motorcycling's answer to Billy Martin. Donning the armor against wind and concrete — helmet, boots, jacket, and gloves — one can't tell president from priest or pauper. All are joined as one under the comfort of cow skin.

Since the first Indian cycle was finished by Hedstrom and Hendee in 1901, motorcycling has at different times played a large role in the maintenance of our country. Don't forget that the military rode Indian and Harley cycles to victory in both world wars and that local law enforcement agencies' orders helped keep both Indian and Harley afloat during the Depression. Todays' officers still use motorcycles for traffic control. (Has anyone watched reruns of CHiPS lately?)

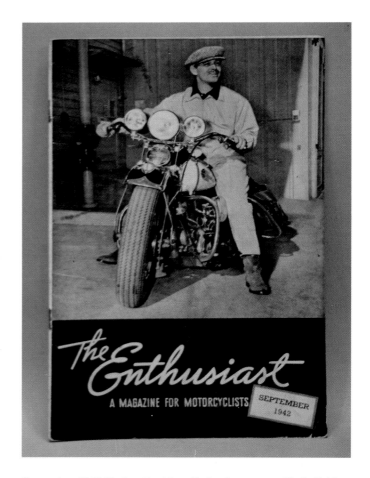

September 1942 Harley Davidson Enthusiast, cover: Clark Gable on dressed out Harley, Gene Autry in centerfold. *Courtesy of the Dunbar Moonlighter Collection.* $15-$50 each

Particularly from the era stretching from the end of World War I to the beginning of World War II, three wheeled cycles carried the commerce of the United States in their specially made compartments. Called servi-cars and package trucks, these cycles delivered mail, groceries, newspapers and parcels, and could usually do so more cheaply and efficiently than their four wheeled counterparts.

Harley 1/4 Ton Package Truck Foldout Sales Brochure, red, white & black, 22" W x 27" W. *Courtesy of the Dunbar Moonlighter Collection.* $25-$75

And don't forget the group of racers who spent many of their waking hours wading through muddy enduro runs, fighting up steep hill-climbs or zipping around board tracks, winning championships amongst the bruises and broken bones that can occur when one messes too much with Mother Nature.

In the golden age of motoring, motorcycles were considered to be the cheaper, hipper, and more exciting way of getting around. Come to think of it, that's largely how they are viewed today. With yearly meets in Daytona, Laconia and Sturgis, motorcyclists still pay annual homage to themselves and their bikes, and have a helluva time doing it.

But, after the glow and the windy flush of last trip have subsided, what's left to remind one of hours in the seat besides a pleasant saddle soreness? Many motorcycling enthusiasts choose to surround themselves with the artifacts of the ninety-five year history of American cycling — sales literature, advertising posters, magazines, books, clothes, awards, pennants, and dealer giveaway trinkets, to name a few categories. Each collection is a personalized exploration into some aspect of biking. Each collector has their own special interests. For some it's collecting the earliest of the early, FAM awards and memorabilia. Others may create shrines to Harley or explore one or more of the short lived companies that prospered in the early part of the century, then fell by the wayside. Still others may just want license plates or magazines. The great thing is that, just like bikers themselves, no two collections are the same.

This book was created as a reference to help guide both potential and advanced collectors, offering photos and descriptions of some of the items that all levels of collectors can hope to put into their own personal collections. This guide should provide enough information for the collector to understand what is easily available and what is scarce.

About the pricing in this book — In creating a price guide for the objects shown in this book, I concentrated on a range. The low end of the range is for an item in fair condition, the high end for an item in mint condition. You may be able to find some of these items for less money. You may have to pay more for others. The fact is that no price guide is 100% accurate because markets change too quickly and prices can be different in different parts of the country. If you paid less than the price range of an item, congratulations. If you paid more, you were not necessarily wrong, you may just be ahead of the market and your piece may catch up or move ahead very soon.

Hubley Hill Climber, cast iron, 6 3/4" long.

Some items shown in this book are relatively easy to obtain, like many of the *Harley Davidson Enthusiasts,* a magazine created by Harley-Davidson in 1916 to both inform its owners about what other Harley owners were doing and, of course, to showcase the latest models coming out on the market. It is still published today. The *Enthusiast* has promoted motorcycling as a sport for all ages — good, clean, wholesome fun. This is an attitude that the mainstream media has rejected. One might expect Jerry Mathers on a 1950s cover as much as Elvis. Some vintage issues of the *Enthusiast* can be bought for as little as a few dollars. Magazines like this are good for collectors, giving them some idea of what riding habits, attitudes, and clothing styles were forty or fifty years ago.

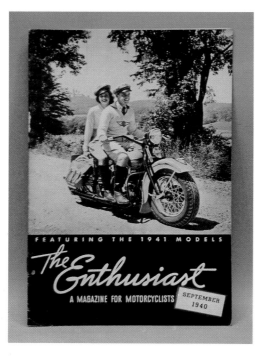

September 1940 Harley Davidson Enthusiast, cover: new Harley, full dressed couple, featuring new 1941 models. *Courtesy of the Dunbar Moonlighter Collection.* $15-$50

As one can imagine, some of the pieces — like the early Indian and Harley posters or FAM items — are very difficult to find and very expensive to buy when someone wants to sell them. A number of readers will gasp when they look over the cast iron toy prices, some of which rival prices of a good used (or even new) bike. The fact is that serious toy collectors have paid, and will continue to pay, a premium for great examples to add to their cast iron toy collections. Anyone who reads this book has to remember, however, that condition is king. Just because someone has a cast iron toy does not mean it has value. The most expensive cast iron toys are the biggest, with the most original accessories, and with the most paint. A lot of people make the mistake of thinking that just having the item constitutes a great value. Not true. Consequently, one cannot go wrong with a great piece in great condition. It may be expensive now, but it's a home run in the long run.

The heyday of American motorcycling is considered to be the period from 1905 to World War I. At that time, more than 300 firms made motorcycles in the United States. Names like Thor, Flying Merkel, Excelsior, Pope, Ace, MM, and Wagner challenged — and occasionally beat — Indian and Harley-Davidson on the race track and the sales floor. Less expensive than automobiles to purchase and run while offering a way off the farm, motorcycle mania hit the United States a decade after bicycling mania. And just like the bicycling fad, after a grand flourish which brought a number of firms into the field, the motorcycle manufacturing industry was primed for a major fallout.

Henry Ford helped the decline by devising the automotive assembly line, thereby making his Model T's more standardized and cheaper than a Harley V-Twin or an Indian Big Chief. Whereas in the 1920s flappers and businessmen were sipping champagne and dancing in speakeasies, Indian and Harley-Davidson were scrambling for ways to sell more cycles, both domestically and abroad. Most of the other makers fell silently by the side of the road.

Hubley Indian Cast Iron Traffic Car, ca. 1930s, 12" Long, removable driver. Real life traffic cars expedited equipment and lights to organize heavy traffic areas. *Courtesy of the Dunbar Moonlight Kid Collection.* $2200-$7500

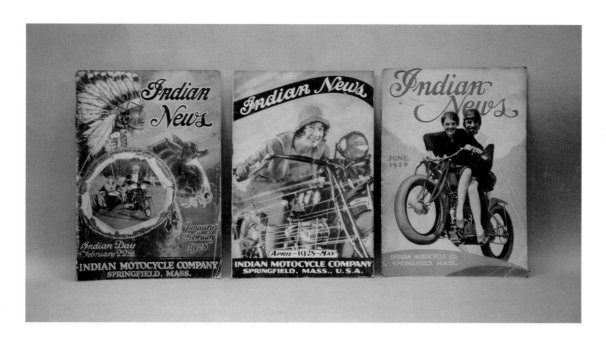

Indian News Magazines, January/February, April/May, & June 1928.
Courtesy of the Dunbar Moonlighter Collection. $20-$75 each

The 1927 Indian Scout 45 Sales Brochure, red, white & black, blowup of Scout 45 & specs. *Courtesy of the Dunbar Moonlighter Collection.* $50-$150

Both companies survived the Depression and World War II, largely through worldwide sales and sales to the military and police departments. But the Indian Motorcycle Company, after a long run, raised the white flag in the early 1950s, leaving Harley-Davidson, as the lone surviving American motorcycle manufacturer, to make the ride solo.

Harley-Davidson spent the 1950s and '60s fighting less expensive imports, as well as a new public image of the biker as outlaw. While the majority of motorcyclists rode for the entertainment and excitement of exploring the countryside, the 1947 Hollister, California riots, Marlon Brando's acting in The Wild Ones (and please note that he rode a Triumph, ahem) and the evolution of Hells Angels all conspired to place the media persona of the biker somewhere between Fonzie and Jack the Ripper.

Surprisingly, the 1970s buy-out of Harley-Davidson by AMF, yes the bowling ball company, helped by infusing H-D with a great amount of capital. Led by the designing efforts of William G. (Willie G.) Davidson and a whole new marketing strategy, H-D climbed back to the top of the motorcycling world and was bought back by a group of H-D executives in 1981. Fifteen years later, one has to look no further than the Harley-Davidson Cafe in New York to see that H-D has the same cachet as Perrier while still appealing to the Budweiser crowd. After all, Malcolm Forbes rode a Harley, and probably still does, while ordering Dom Perignon from that big biker bar in the sky.

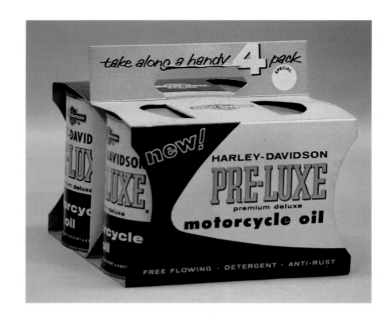

Harley-Davidson Pre-Luxe Motorcycle Oil four pack.

Harley-Davidson's winged logo is now the symbol of American motorcycling, sought after by the rich and rebels alike, all happily rolling down one road. Because of H-D's longevity, it is an obvious conclusion that most of the collectibles available are from that company. Not far behind Harley, one will find a number of Indian motorcycle collectibles. The memory of the Springfield, Massachusetts company has been kept alive by Charlie and Esta Mantos, through their effort and passion, culminating in the Indian Motocycle Museum. Much of the original Indian archival material resides under their roof, surrounded by vintage cycles, toys, racing trophies, and photos of past champions — a must see for anyone who comes to Springfield, Massachusetts.

Harley Davidson Patch, winged logo, 10" Long. *Courtesy of the Dunbar Moonlighter Collection.* $50-$100

Unfortunately, since many of the other companies only produced small numbers of bikes for a limited time, the amount of material available is likewise small. Companies like Excelsior-Henderson were sold out, consolidated, and then eventually fazed out. Iver Johnson, from Massachusetts, also made guns and bicycles. Sears, Schwinn, and Columbia all endeavored to produce motorcycles as a logical progression from bicycles, then found the division unprofitable and returned to their better selling products. Therefore, any early material in excellent condition from short-lived companies demands attention.

Iver Johnson Arms & Cycle Works 1916 Sales Book & Price List of Bicycles, Motorcycles & Firearms, with a color panel wihin the catalogue of W.W. I soldiers on bicycles, 80 pages total, 20 pages devoted to pictures & specs of motorcycle line, 4 1/2" W x 7" H, rare catalogue. *Courtesy of the Dunbar Moonlighter Collection.* $300-$600

Motorcycle Illustrated Magazine Oct 19, 1911, Indian Solo on cover. *Courtesy of the Dunbar Moonlighter Collection.* $100-$350

How to collect: In collecting motorcycle memorabilia, or motomobilia as some are now calling this field, one has to remember basic tenets in hunting down, acquiring, and enjoying one's newfound treasures. This philosophy has been honed in twenty years of collecting and dealing in antique toys, advertising, automobilia, motomobilia and Americana:

1) Buy what you like. If you are buying to collect, get what you want. You are the one who is going to look at it for a long time. It is an extension of your personality. Don't just buy something if someone tells you it's a good deal, unless you plan to put it away as an investment and sell it later to finance a piece that you really want for yourself. Collectors collect because what they're after feeds a void in their soul that cannot be found in their job, just like any hobby. So don't cheat yourself. Many people start out buying cheap because they don't know very much and they're afraid to make mistakes. But, if you buy something that truly speaks to you and makes you feel whole, even if you spend a few extra bucks, what does it matter in the long run?

2) Find dealers with whom you can develop a good rapport, ones that you can trust, and ones that will guarantee their items in good faith, in written form if necessary. The collectibles market in general has always been a casual market. You buy from dealers and other collectors on faith and trust in them for their own accrued experience, knowledge, and expertise. But, as prices go up in any category and there is more money to be made, reproductions and forgeries will start to appear. Make dealers answer your questions and back up their items. Good ones will because they want to you to be satisfied and they want you to come back and deal with them again.

3) Buy items in the best condition that you can afford. Many times over the years I've heard the phrase, either in buying or selling, "That price is too high." Usually a price is high because the item is either the best or one of the best examples that you can find in the best condition. Those are always the pieces that appreciate the most over time. Those are always the pieces that are fought over in auctions because of the simple facts of supply versus demand. You can't make any more of these items from the 1920s, '30s, etc., so you have to get what you can when you can. The opportunities to buy rare stuff in great condition are limited. Don't let a high price tag dissuade you if the condition is spectacular and the item is scarce. Conversely, think about what you are buying and don't let yourself be pushed into something if you don't absolutely love it and don't feel comfortable with it. We all buy on emotion and that can work both ways. Take your time, learn as much as you can, and enjoy what you get.

Where can I find this stuff? That's a tricky question to answer. In many parts of the country, particularly in the Northeast and California, there are semi-frequent motorcycling related shows, especially during the summer months. At the large motorcycling meets, such as the annual American Motorcyclist Association (AMA) meet in Westerville, Ohio, there are usually shows where you can buy or sell pieces while meeting other collectors and dealers.

To find out about shows, dealers, and collectors, you can start by reading one of the many motorcycling related publications on the market, such the "Motorcycle Parts And Literature For Sale" section of *Hemmings Motor News*. Other magazines include *Motorcycle Shopper Magazine, Walneck's Classic Cycle-Trader, The Antique Motorcycle, American Motorcyclist* published by the AMA, *Motorcycle Collector,* and *Mobilia* (see the list at the end of this book).

Once you start meeting people, it is natural to learn the network that exists across the country. It takes time, some study, mileage, and phone calls, but you will soon learn the ins-and-outs of the collecting game.

Note that this book is covering only American related motorcycling collectibles. I felt that trying to include every biking collectible from Ducati to Honda would be difficult and would end up in a book lacking any depth. By limiting the scope of collectibles to those about biking in the United States, I think the book will give collectors a better feeling for the history of motorcycling in their own back yard.

AMA Gypsy Tour Award Belt Buckle of 1958, AMA Logo, 2 1/2" W x 1 3/4" H. 1958 AMA Gypsy Tour Award Badge. *Courtesy of the Dunbar Moonlighter Collection.* $50-$125.

SALES LITERATURE

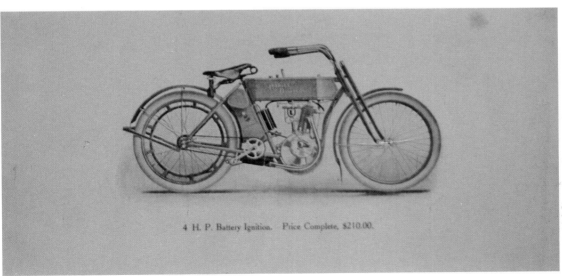

CHAPTER I

4 H. P. Battery Ignition.   Price Complete, $210.00.

A 1910 Harley Davidson Motorcycle Sales Catalogue. *Courtesy of the Doug Leikala Collection.* $200-$650

America's golden advertising age of lithography extended loosely from the 1870s to the 1920s. The most graphic and beautiful motorcycling posters, sales brochures, and booklets were created from the early 1900s to the 1920s. Boldly painted posters decorated dealers' walls and jump-started potential riders' imaginations, taking them to here-to-fore impossible destinations. Handouts were one of dealers' major methods of showing off their wares to a public not yet listening to the radio or watching television. Whether companies shot for the excitement and speed of motorcycling on the edge from atop an Indian, or they opted for the wonderment of riding on their Pope and enjoying a waterfall, the magnificent colors and artistry some of these pieces have made them works of art in their own right.

Later literature would use photos in place of art work. These handbills also have a value; however, in general the worth is substantially less than their more ornate predecessors.

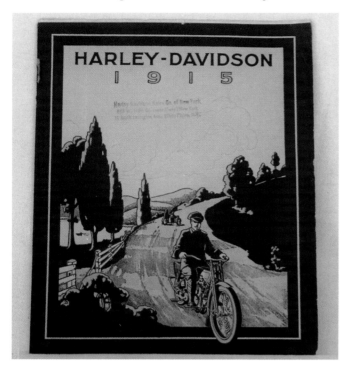

The 1915 Harley Davidson Sales Catalogue on all 1915 models. *Courtesy of Ken Kalustian.* $100-$500

A 1919 Harley Davidson Sales Catalogue, W.W. I. *Courtesy of the Doug Leikala Collection.* $100-$500

Harley Davidson 1915, 1916, and 1917 Catalogues. *Courtesy of Cris & Pat Simmons.* $100-$500 each

Six catalogues: 1912 Big 5 Sales Catalogue, Minneapolis Motorcycle Company; 1911 The Flying Merkel Sales Catalogue; 1913 Dayton Motorcycle Catalogue; 1911, 1912, and 1911 Thor Motorcycle Catalogues, *Courtesy of Cris & Pat Simmons.* $100-$800 each

The Story Of The 1907 Indian Catalogue, the 1913 Indian Motocycles Catalogue. A 1920 Indian Motorcycle Catalogue, The Gentleman's Mount. A 1920 Indian Side Car Catalogue. *Courtesy of Cris & Pat Simmons.* $200-$700 each

Left to right, top to bottom: 1911, 1917, 1915, 1912 Indian Motorcycle Catalogues. *Courtesy of Cris & Pat Simmons.* $100-$700 each

A 1917 Reading Standard Catalogue. *Courtesy of Cris & Pat Simmons.* $200-$600

Reading Standard Catalogues, ca. 1910-1917. *Courtesy of Cris & Pat Simmons.* $100-$600 each

A 1911 American Motorcycles Sales Catalogue and a 1915 Dayton Motorycles Sales Catalogue. *Courtesy of Cris & Pat Simmons.* $100-$600 each

A pair of Emblem Motorcycle And Bicycles Catalogues, ca. 1910. The Flying Merkel Sales Catalogue, ca. 1912. *Courtesy of Cris & Pat Simmons.* $100-$600 each

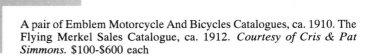

Left to right, top to bottom: Curtiss Motorcycle Sales Catalogue, 1904; The Light Sales Catalogue, ca. 1910; Excelsior Instruction Book, ca. 1915; New Era Auto Cycle Sales Catalogue, 1909; and a Peerless Motorcycles Sales Catalogues, 1911. *Courtesy of Cris & Pat Simmons.* Catalogues $100-$600 each, Instruction Book, $50-$200

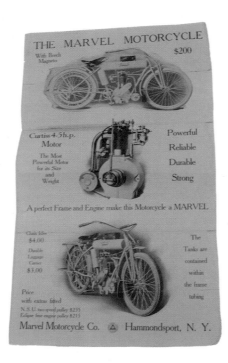

A 1912 Marvel Motorcycle Sales Brochure. *Courtesy of Cris & Pat Simmons.* $100-$350

Henderson 1916, 1917 Sales Catalogues. *Courtesy of Cris & Pat Simmons.* $100-$500

Left to right, top to bottom: 1926 Harley Rider's Handbook, $50-$150; 1927 Sales Brochure, $50-$200; 1912 Parts Catalogue, $50-$250; 1920s Accessories Catalogue, $50-$150; 1920s Brochure, $50-$200; and a 1919 Rider's Hand Book, $50-$150. *Courtesy of Cris & Pat Simmons.*

Iver Johnson Arms & Cycle Works 1916 Sales Book & Price List of Bicycles, Motorcycles & Firearms, with a color panel wihin the catalogue of W.W. I soldiers on bicycles, 80 pages total, 20 pages devoted to pictures & specs of motorcycle line, 4 1/2" W x 7" H, rare catalogue. *Courtesy of the Dunbar Moonlighter Collection.* $300-$600

The 1913 Pope Motorcycles Catalogue, Westfield, Massachusetts. *Courtesy of Cris & Pat Simmons.* $200-$800

A 1915 Pope Motorcycles Catalogue, Westfield, Massachusetts. *Courtesy of Cris & Pat Simmons.* $200-$600

Ace Motorcycle Catalogue, 1920. *Courtesy of Cris & Pat Simmons.*
$200-$600

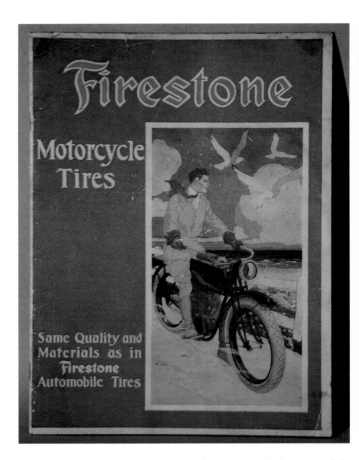

Firestone Motorcycle Tires Sales Brochure, ca. 1915. *Courtesy of the Doug Leikala Collection.* $50-$250

Dayton Motorcycle Framed Foldout Sales Brochure, ca. 1915. *Courtesy of the Dunbar Moonlighter Collection.* $100-$350

The 1927 Excelsior & Henderson Color Sales Brochure featuring Super Sport, Super X, Henderson Models, 10 pages. *Courtesy of the Dunbar Moonlighter Collection.* $100-$250

Cleveland Cycles 1926 Foldout Sales Brochure, centerfold of Cleveland "Four", 9" W x 12" H, folds out to 18" W x 12" H. *Courtesy of the Dunbar Moonlighter Collection.* $150-$350

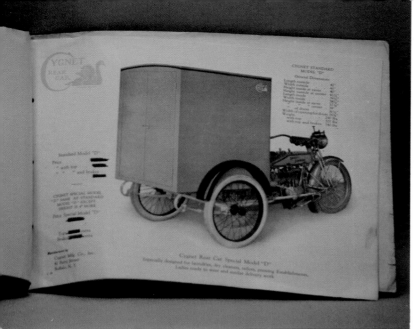

Cygnet Rear Car Sales Catalogue, featuring attachments turning motorcycles into multi-person vehicles, package trucks & all weather vehicles, 8 different attachment bodies as shown on 11" W x 8" H photos of Harleys & Indians, with specs & prices, with 2 separate testimonial letters, order form, 12" W x 9" overall size. *Courtesy of the Dunbar Moonlighter Collection.* $200-$400

A 1925 Indian Motorcycles 8 Page Sales Booklet, red, white & black, featuring blowups of Prince, Scout, and Chief with Princess Sidecar. *Courtesy of the Dunbar Moonlighter Collection.* $75-$200

Indian 25th Anniversary Sales Booklet, 1926, red, black, & white, featuring Indian Servicars, such as Fire Patrol, blowups of Prince, Scout, Chief, and Big Chief with Princess Sidecar. *Courtesy of the Dunbar Moonlighter Collection.* $75-$200

A 1927 Indian Foldout Sales Brochure, featuring latest Prince, Scout, Chief, and Big Chief Models in red, white & black. *Courtesy of the Dunbar Moonlighter Collection.* $75-$200

The 1927 Indian Scout 45 Sales Brochure, red, white & black, blowup of Scout 45 & specs. *Courtesy of the Dunbar Moonlighter Collection.* $50-$150

Indian Motorcycle Foldout Sales Brochure, 1928, for Servicar, black & white & red. *Courtesy of the Dunbar Moonlighter Collection.* $25-$50

The 1927 Indian Sales 19 Page Booklet, red, white & black, featuring 1927 models, pictures of racing stars Johnnie Seymour, Orie Steele, and Eddie Mitchell. *Courtesy of the Dunbar Moonlighter Collection.* $75-$200

Indian Foldout Sales Brochures, six panels, the 1925 Prince, $75-$150; the 1925 Indian Scout, $50-$150; and the 1928 Indian Motorcycle Sales Sheet. $10-$40 *Courtesy of the Dunbar Moonlighter Collection.*

A 1923 Harley Sales Brochure. *Courtesy of the Dunbar Moonlighter Collection.* $50-$250

The 1928 Indian Scout Sales Brochure, "Power-Economy-Swiftness-Comfort-Safety-Stamina," with hillclimb stats, AMA records. $50-$100. The 1928 Indian Foldout Sales Brochure For Safety. *Courtesy of the Dunbar Moonlighter Collection.* $10-$35

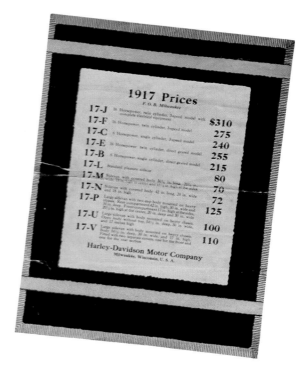

A 1917 Harley Davidson Fold Out Sales Brochure, featuring "Power & Speed," with prices, 9" W x 12" H, folds out To 18" W x 12" H. *Courtesy of the Dunbar Moonlighter Collection.* $50-$150

Harley Sales Catalogues, 1922 & 1929. *Courtesy of the Doug Leikala Collection.* $50-$250 each

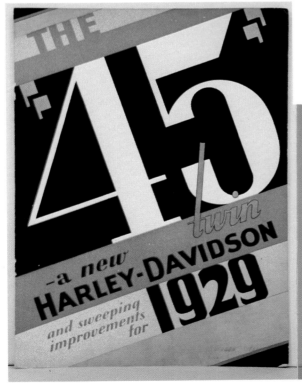

A 1929 Harley Davidson Foldout Sales Brochure, featuring "45" Twin, Deco colors, 9" W x 12" H, folds out to 18" W x 12" H. *Courtesy of the Dunbar Moonlighter Collection.* $50-$150

The 1931 Harley Davidson Police Motorcycle Sales Brochure, 17" W x 11" H. *Courtesy of the Dunbar Moonlighter Collection.* $100-$300

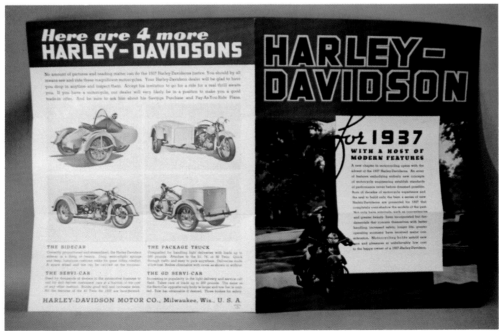

Harley Davidson Sales Brochure for 1937. *Courtesy of the Doug Leikala Collection.* $50-$125

A 1927 Harley Davidson Fold Out Sales Brochure, featuring line of package truck cycles, photos of fleets used from newspaper delivery and Fannie Farmer Candies to mail in Norway, 16" W x 17" H folded out, very nice condition. *Courtesy of the Dunbar Moonlighter Collection.* $50-$150

Harley Davidson Sales Brochure for 1941. *Courtesy of the Doug Leikala Collection.* $50-$150

Harley Davidson Fold Out Sales Brochure, 1947. *Courtesy of the Doug Leikala Collection.* $25-$100

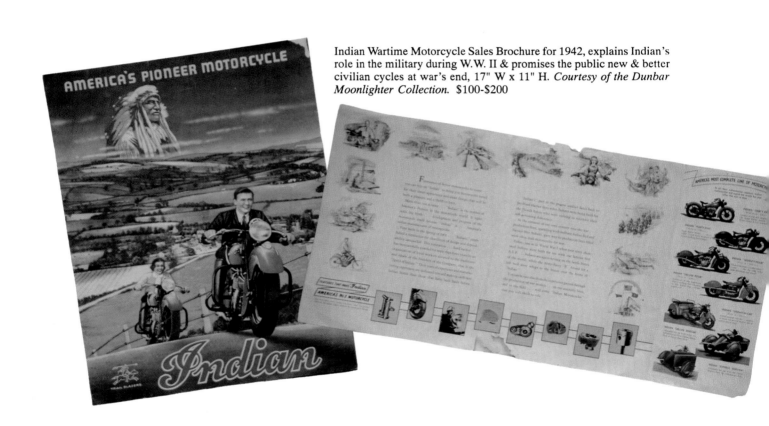

Indian Wartime Motorcycle Sales Brochure for 1942, explains Indian's role in the military during W.W. II & promises the public new & better civilian cycles at war's end, 17" W x 11" H. *Courtesy of the Dunbar Moonlighter Collection.* $100-$200

Indian Catalogues, 1929-'41. *Courtesy of the Dunbar Moonlighter Collection.* $50-$125 each

Harley Davidson 1952 16 Page Sales Booklet, featuring specs & pictures of 1952 models, including "74," "61," "45," "125," with testimonials, photos of bikes in action, 9" W x 6" H. *Courtesy of the Dunbar Moonlighter Collection.* $25-$100

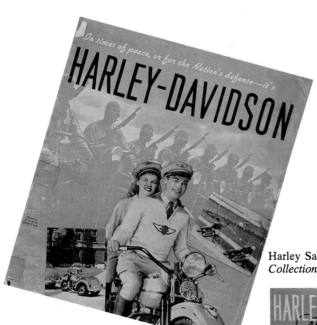

Harley Sales Brochure, 1950s. *Courtesy of the Dunbar Moonlighter Collection.* $50-$100

Harley 1/4 Ton Package Truck Foldout Sales Brochure, red, white & black, 22" W x 27" W. *Courtesy of the Dunbar Moonlighter Collection.* $25-$75

Harley 125 Double Sided Sales Brochure, ca. 1950, 7" W x 10" H. *Courtesy of the Dunbar Moonlighter Collection.* $20-$40

Harley Davidson Fold Out Sales Brochures for 1954, 1963, 1959. *Courtesy of the Doug Leikala Collection.* $25-$100 each

Harley Package Truck Foldout Sales Brochure for 1959, featuring specs, picture of cycle, 21" W x 22" H. Framed. *Courtesy of the Dunbar Moonlighter Collection.* $25-$75

A 1956 Harley Davidson Hydra Glide Framed Foldout Sales Brochure, rose, white, & black, with specs & picture of cycle, 21" W x 15" H. *Courtesy of the Dunbar Moonlighter Collection.* $25-$75

ADVERTISING

CHAPTER II

Oliver's
TRANSPORTATION MEMORABILIA AUCTION

HARLEY-DAVIDSON

SATURDAY, MARCH 9, 1991 AT 10:00 AM

Harley Davidson Hillclimber Poster, ca. 1928, 40" W x 57" H. *Courtesy of the Dunbar Moonlight Kid Collection.* $3000-$12000

Collectors not only seek out early paper advertising, but also want neon signs and clocks to light up their garages or playrooms. Harley and Indian neon signs are considered top-of-the-line. Motor oil cans are sought after by both motorcycling fans and oil can collectors, from the earliest jars and tins, to the later quart cans. Through their dealerships Indian, Harley, and Excelsior all sold oil in their own glass jars or tin cans; other oil companies, such as Oilzum, Texaco, etc., also made oil specifically for motorcycles. As with literature and posters, the earlier and more graphic the can, the more valuable and desirable it will be.

Indian Scout Poster With Male Rider, ca. 1920s, 36" W x 24" H. *Courtesy of Ron Pritchard.* $2000-$6500

The Indian Motocycle Poster With Lady Rider, ca. 1915, 14" W x 24"
H. *Courtesy of Ron Pritchard.* $1500-$6500

Indian Motocycles Fan With Crazy Indian, ca. 1915. *Courtesy of the Doug Leikala Collection.* $200-$800

Thomas Motorcycles Ad, ca. 1910. *Courtesy of the Doug Leikala Collection.* $75-$200

Registered FAM Repair Shop Porcelain Sign, white with cobalt blue, 1912-1919. *Courtesy of the Doug Leikala Collection.* $1000-$3500

Henderson Motorcycle Storecard, Colson's Motorcycle Garage, ca. 1920. *Courtesy of the Doug Leikala Collection.* Beware of reproductions. $75-$150

Coke is it — 1930 Harley Davidson Package Truck Delivering Coca Cola. *Courtesy of the Doug Leikala Collection.* $35-$75

Schaber's Cycle Shop Calendar, Ithaca, New York, 1938, Harley
Davidson motorcycle dealer. *Courtesy of the Doug Leikala Collection.* $100-$250

Drive Safely, 20th Century Taxi Thermometer, cop on cycle, 1930s.
Courtesy of the Doug Leikala Collection. $200-$650

Schaber's Motorcycle Shop 1932 Calendar, Harley Davidson motorcycle dealer. *Courtesy of the Doug Leikala Collection.* $150-$350

Harley Davidson Motorcycles Shop Valance For Around Windows
Decal, ca. 1930s. *Courtesy of the Doug Leikala Collection.* $100-$200

Indian Motorcycles Neon Sign, ca. 1930s-1940s. *Courtesy of the Doug
Leikala Collection.* $500-$2500

Harley Davidson Neon Sign For Indoors, ca. 1930-40, hung in store window or above sales counter. *Courtesy of Ken Kalustian.* $2000-$4000

Harley Davidson Neon Clock, ca. 1930s-1940s. This clock came from a Harley Davidson dealership in Pueblo, Colorado. *Courtesy of Ken Kalustian,* $1000-$3000

Harley Davidson Light Up Clock, ca. 1949. *Courtesy of the Doug Leikala Collection.* $1000-$3000

Cushman Sales · Service Motor Scooter Tin Sign, ca. 1930s. *Courtesy of the Doug Leikala Collection.* $150-$500

Harley Davidson Cycles Tin Sign, Parts, Service, Motorcycles, Servi-Cars, ca. 1930s. *Courtesy of the Doug Leikala Collection.* $400-$900

Harley Davidson Motorcycles Tin Sign, Dishong Cycle Shop, Everett, Pennsylvania, Sales and Service, ca. 1940s. *Courtesy of the Doug Leikala Collection.* $100-$400

Indian Motorcycles Light Up Clock, ca. 1940s. *Courtesy of the Doug Leikala Collection.* $200-$650

Harley Davidson Motorcycles Lightup Sign, ca. late 1950s-early 1960s. *Courtesy of the Doug Leikala Collection.* $200-$650

Indian Alarm Clock, ca. 1930s-1940s. *Courtesy of the Doug Leikala Collection.* $50-$150

Harley Davidson Motorcycles Light Up Sign, ca. late 1950s-early 1960s. *Courtesy of the Doug Leikala Collection.* $200-$650

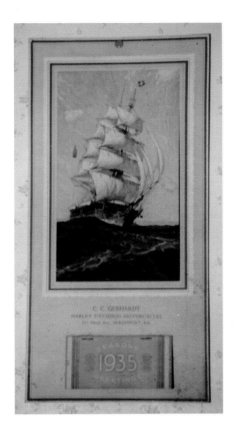

S. C. Gebhardt Motorcycles & Bicycles 1935 Calendar. *Courtesy of the Doug Leikala Collection.* $75-$125

S. C. Gebhardt Motorcycles & Bicycles 1937 Calendar. *Courtesy of the Doug Leikala Collection.* $75-$125

The 1936 S. C. Gebhart Harley Davidson Motorcycles Calendar, McKeesport, Pennsylvania, village cottage scene. *Courtesy of the Doug Leikala Collection.* $50-$100

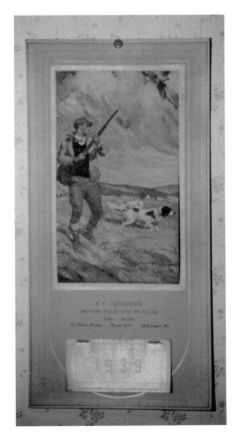

S. C. Gebhardt Motorcycles & Bicycles 1939 Calendar. *Courtesy of the Doug Leikala Collection.* $75-$125

Set Of 3 Harley Davidson Parts Ads, Spark Plugs, Gunk, Pre-Luxe Motorcycle Oil, ca. 1950s. *Courtesy of the Doug Leikala Collection.* $30-$75 each

Oilzum Motorcycle Oil Tin Sign, "Choice Of Champions!", ca. 1950s. *Courtesy of the Doug Leikala Collection.* $75-$300

Harley Davidson/Gunk Thermometer, ca. 1950s. *Courtesy of the Doug Leikala Collection.* $200-$700

Gunk Neon Sign, Harley Davidson Motorcycle Cleaner, ca. 1950s. *Courtesy of the Doug Leikala Collection.* $100-$450

A 1958 Harley-Davidson Parts & Accessories Framed Banner. *Courtesy of the Doug Leikala Collection.* $200-$1200

"We Service What We Sell" Harley Davidson Parts Banner, 1960s. *Courtesy of the Doug Leikala Collection.* $50-$150

"This Man May Save Your Life!", America Fare Insurance Group Ad, framed, ca. 1950s. *Courtesy of the Doug Leikala Collection.* $100-$250

Ad For Deluxe Plastic Handle Bar Grips For Motorcycles, 1950s. *Courtesy of the Doug Leikala Collection.* $50-$125

Harley Cigarettes Tin Sign, ca. 1980s. Brand was discounted soon after manufactured. This is one of a small number of signs that were made but never circulated. *Courtesy of the Dunbar Moonlighter Collection.* $150-$350

Package of Harley Davidson Cigarettes, ca. 1980s. *Courtesy of the Doug Leikala Collection.* $2-$8

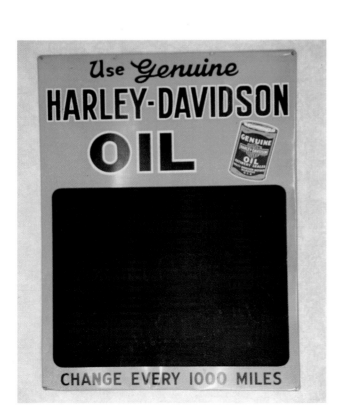

Use Genuine Harley Davidson Oil Tin Chalkboard, ca. 1940s. *Courtesy of the Doug Leikala Collection.* $100-$450

Early Can Harley Davidson Oil 5 Gallon Can, ca. 1909. *Courtesy of the Doug Leikala Collection.* $500-$1500

Harley Davidson 5 Gallon Motor Oil Can, ca. 1920s. *Courtesy of the Doug Leikala Collection.* $300-$1000

Indian Top Lube, Harley Davidson Oil For 4-Cycle Motors, Indian
Transmission Oil, Indian Motorcycle Oil, Indian Penetrating Oil, ca.
1940s-50s. *Courtesy of Ken Kalustian.* $25-$150

Harley Davidson Quart Racing Oil, 60 Wt. $75-$200. Indian Oil Can,
$50-$125. Oilzum Motorcycle Oil Can. *Courtesy of Ken Kalustian.*
$100-$300.

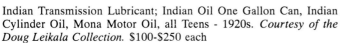

Indian Transmission Lubricant; Indian Oil One Gallon Can, Indian
Cylinder Oil, Mona Motor Oil, all Teens - 1920s. *Courtesy of the
Doug Leikala Collection.* $100-$250 each

Harley Davidson Motorcycle Oil Match Sticks, 1950s, $50-$125;
Motorcycling Lighter, 1950s, $75-$200; Harley Lighter, 1950s, $100-
$300; Harley Lighter with Cycle, 1960s, $75-$200; Quart Harley
Davidson Motorcycle Oil Match Sticks, 1950s; *All Courtesy of the
Doug Leikala Collection.* $75-$125.

Can Harley Davidson Racing Motorcycle Oil, ca. 1960s. $25-$75.
Harley Davidson Gunk Can, 1950-60. *Courtesy of Ken Kalustian.* $25-
$85

Gallon Genuine Harley Davidson Motor Oil Can ca. 1930s, $300-$700;
Genuine Harley Davidson Motor Oil Jar, ca. 1940s. $50-$200 *Courtesy of the Doug Leikala Collection.*

Harley- Davidson Motorcycle & Sidecar Framed Ad, 1919. *Courtesy of the Doug Leikala Collection.* $100-$300

BOOKS & MAGAZINES

Motorcycle Illustrated Magazine Sept 15, 1908, Indian Solo on cover with racing driver. *Courtesy of the Dunbar Moonlighter Collection.* $100-$350

Motorcycling books and magazines appeared almost as soon as the first two-wheeled rumble could be heard at the turn of the century. In 1912, Andrew Carey Lincoln started a series called *Motorcycle Chums* that took friends Budge and Alec all about the countryside, in taverns, up and down mountains, even through the woods and right past Grandma's house if the clutch wasn't working correctly. Other authors followed suit and entertained a whole generation with trailside exploits. These books used to be found for as little as 50 cents or a dollar. Now they sell for anywhere between $10 and $50, depending on condition and the knowledge of the vendor.

Motorcycle Illustrated magazines premiered in the early 1900s. Their pages were full of names now in the motorcycle graveyard: Ace, Yale, New Era, Racycle, Ideal, Emblem, etc. Early covers featured Indian bikes and advertising, as they were considered top-of-the-heap until Harley overtook them in the 1920s. In 1916, to inspire consumer loyalty and to inform Harley owners of the latest and greatest Harley feats, Harley-Davidson created the *Enthusiast*, now the longest continually running motorcycle publication in the country.

Produced in enough quantity to still be found today, many collectors make it a goal to acquire every copy of the *Enthusiast* run, an almost impossible feat. However, collecting these magazines is a very good way to enter the hobby. Many copies can be bought very reasonably, and they carry a lot of information about the sport of motorcycling. Issues will carry first person experiences from places around the globe as distant as Africa and as close as Aberdeen. Even when the company did not support Harley racers, the magazine would still carry race results, with competitors' names magically marked out as if they never finished.

Copies of the *Indian News, Honest Injun,* and *Wigwam News* (Indian's owner and in house magazines) tried to emulate the *Enthusiast*, with mixed results. Consequently, they were produced in smaller numbers, are harder to find today, and sell for higher prices. Many of the 1920s covers were far more dynamic and interesting than their competitors. Stunning action art work featured speeding women and wahooing chiefs. In most *Enthusiasts*, sometimes the reader had to settle for sedate sidecar trip photos, showing fishing or spelunking expeditions.

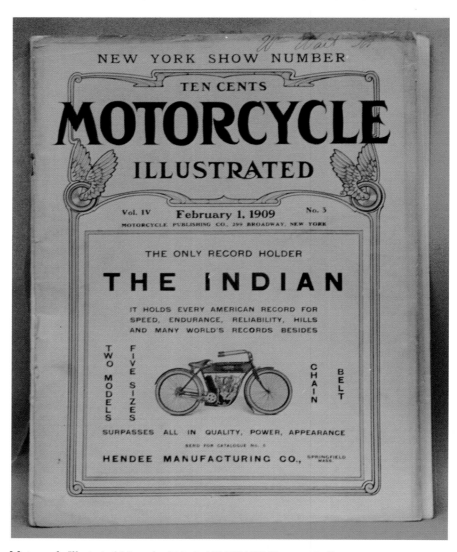

Motorcycle Illustrated Magazine Feb. 1, 1909 Vol IV #3, cover Indian Solo as American Record Holder. *Courtesy of the Dunbar Moonlighter Collection.* $100-$350

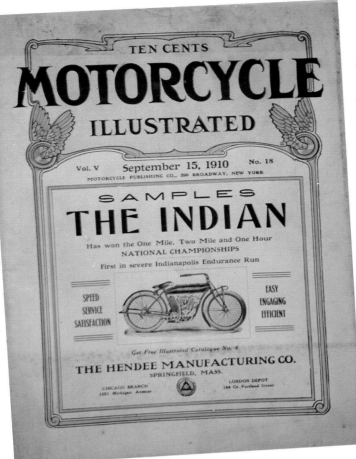

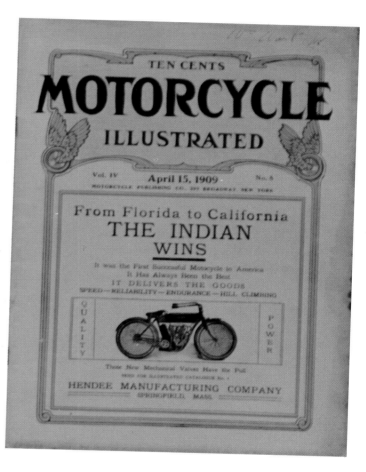

Motorcycle Illustrated Magazine Sept 15, 1910. *Courtesy of the Dunbar Moonlighter Collection.* $100-$350

Motorcycle Illustrated Magazine April 15, 1909 Vol IV #8. *Courtesy of the Dunbar Moonlighter Collection.* $100-$350

Motorcycle Illustrated Magazine, March 15, 1909, Vol IV #6. *Courtesy of the Dunbar Moonlighter Collection.* $100-$350

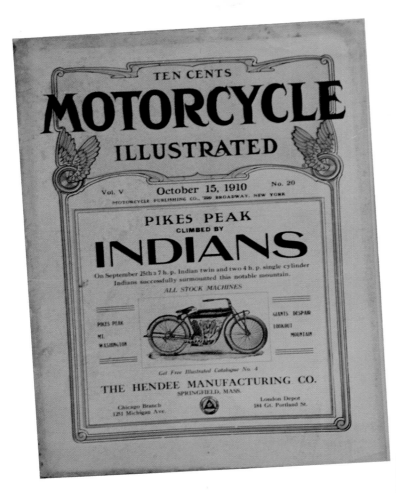

Motorcycle Illustrated Magazine, Oct. 15, 1910, cover "Pikes Peak Climbed By Indians." *Courtesy of the Dunbar Moonlighter Collection.* $100-$350

Motorcycle Illustrated Magazine Oct 19, 1911, Indian Solo on cover. *Courtesy of the Dunbar Moonlighter Collection.* $100-$350

Motorcycle Illustrated Magazine, Jan. 12, 1911. *Courtesy of the Dunbar Moonlighter Collection.* $100-$350

The 1928 Western Motorcyclist & Bicyclist Magazine, featuring color centerfold of Indian "4," also articles & ads for Harley, Whizzer, Excelsior, 50 pages. *Courtesy of the Dunbar Moonlighter Collection.* $25-$100

Motorcycle & Bicycle Illustrated Magazine Aug 14, 1919. *Courtesy of the Dunbar Moonlighter Collection.* $50-$200

The Harley Davidson Enthusiast, August 1916. *Courtesy of the Doug Leikala Collection.* $25-$125

Harley Davidson Enthusiast, ca. 1917-1919. *Courtesy of the Doug Leikala Collection.* $25-$125

Harley Davidson Enthusiasts, 1916 & 1917. *Courtesy of the Doug Leikala Collection.* $25-$125 each

No. 16 & 17 Harley Davidson Enthusiasts, ca. 1917-1919. *Courtesy of the Doug Leikala Collection.* $25-$125

Nos. 18 & 19 Harley Davidson Enthusiasts, ca. 1917-19. *Courtesy of the Doug Leikala Collection.* $25-$125 each

Harley Davidson Enthusiasts, ca. 1918-20. *Courtesy of the Doug Leikala Collection.* $25-$125 each

Harley Davidson Enthusiasts, May, July, August, & Sept. 1920. *Courtesy of the Doug Leikala Collection.* $25-$125 each

Harley Davidson Enthusiasts, October, November, & Dec. 1920. *Courtesy of the Doug Leikala Collection.* $25-$125 each

Harley Davidson Enthusiasts, January, February, March, & April 1921. *Courtesy of the Doug Leikala Collection.* $25-$125 each

Harley Davidson Enthusiasts, May, June, July, & August 1921. *Courtesy of the Doug Leikala Collection.* $25-$125 each

Harley Davidson Enthusiasts, September, October, November, & December 1921. *Courtesy of the Doug Leikala Collection.* $25-$125 each

Harley Davidson Enthusiasts, January, February, March, & April 1922. *Courtesy of the Doug Leikala Collection.* $20-$75 each

Harley Davidson Enthusiasts, May, June, July, & August 1922. *Courtesy of the Doug Leikala Collection.* $20-$75 each

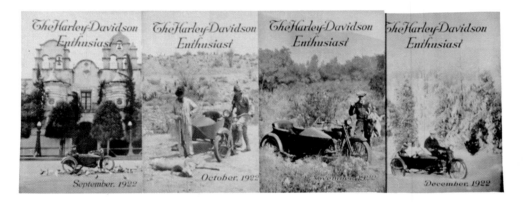

Harley Davidson Enthusiasts, September, October, November, & December 1922. *Courtesy of the Doug Leikala Collection.* $20-$75 each

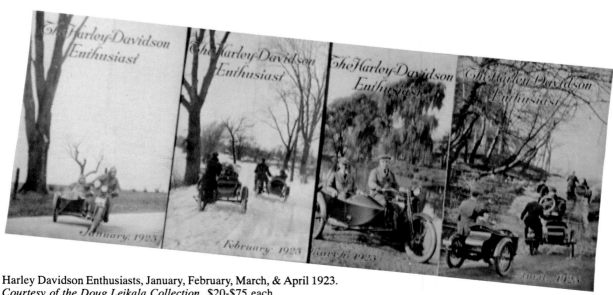

Harley Davidson Enthusiasts, January, February, March, & April 1923. *Courtesy of the Doug Leikala Collection.* $20-$75 each

Harley Davidson Enthusiasts, May, June, July, & August 1923. *Courtesy of the Doug Leikala Collection.* $20-$75 each

Harley Davidson Enthusiasts, September, October, November, & December 1923. *Courtesy of the Doug Leikala Collection.* $20-$75 each

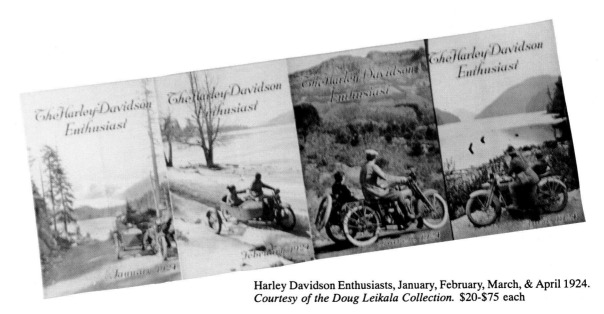

Harley Davidson Enthusiasts, January, February, March, & April 1924. *Courtesy of the Doug Leikala Collection.* $20-$75 each

Harley Davidson Enthusiasts, May, June, July, & August 1924. *Courtesy of the Doug Leikala Collection.* $20-$75 each

Harley Davidson Enthusiasts, September, October, November, & December 1924. *Courtesy of the Doug Leikala Collection.* $20-$75 each

Harley Davidson Enthusiasts, January, February, March, & April 1925. *Courtesy of the Doug Leikala Collection.* $20-$75 each

Nobby Ned Comic From 1924 Enthusiast, found from time to time on the back covers. *Courtesy of the Doug Leikala Collection.*

Harley Davidson Enthusiasts, May, June, July, & August 1925. *Courtesy of the Doug Leikala Collection.* $20-$75 each

Harley Davidson Enthusiasts, September, October, November, & December 1925. *Courtesy of the Doug Leikala Collection.* $20-$75 each

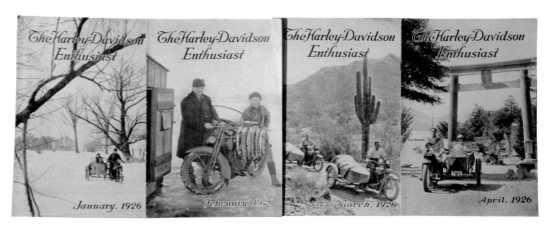

Harley Davidson Enthusiasts, January, February, March, & April 1926. *Courtesy of the Doug Leikala Collection.* $20-$75 each

Harley Davidson Enthusiasts, May, June, July, & August 1926. *Courtesy of the Doug Leikala Collection.* $20-$75 each

Harley Davidson Enthusiasts, September, October, & November-December 1926. *Courtesy of the Doug Leikala Collection.* $20-$75 each

Harley Davidson Enthusiasts, January, February, March, & April 1927. *Courtesy of the Doug Leikala Collection.* $20-$75 each

Harley Davidson Enthusiasts, May, June, July, & August 1927. *Courtesy of the Doug Leikala Collection.* $20-$75 each

Harley Davidson Enthusiasts, September, October, November, & December 1927. *Courtesy of the Doug Leikala Collection.* $20-$75 each

Harley Davidson Enthusiasts, January, February, March, & April 1928. *Courtesy of the Doug Leikala Collection.* $20-$75 each

Harley Davidson Enthusiasts, May, June, July, & August 1928. *Courtesy of the Doug Leikala Collection.* $20-$75 each

Harley Davidson Enthusiasts, September, October, November, & December 1928. *Courtesy of the Doug Leikala Collection.* $20-$75 each

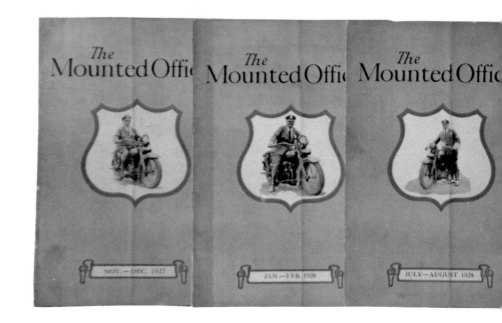

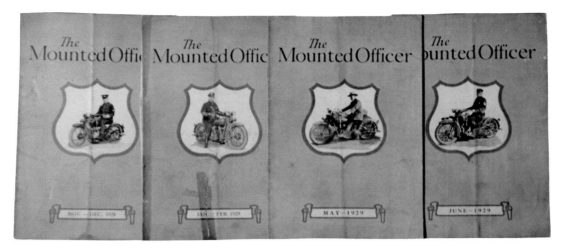

Four Copies Of The Mounted Officer, Nov.-Dec. 1928, Jan.-Feb. 1929,
May 1929, June 1929. *Courtesy of the Doug Leikala Collection.* $20-
$75 each

Three Copies Of The Mounted Officer, July 1929, Sept.-Oct. 1929,
Nov.-Dec. 1929. *Courtesy of the Doug Leikala Collection.* $20-$75
each

Four Copies Of The Mounted Officer, the police companion to the
Harley Davidson Enthusiast magazine. Nov.-Dec. 1927, Jan.-Feb. 1928,
July-August 1928, Sept.-Oct., 1928. *Courtesy of the Doug Leikala
Collection.* $20-$75 each

The Harley Davidson Enthusiast, January, February, March, & April
1929. *Courtesy of the Doug Leikala Collection.* $15-$50 each

Four Copies Of The Harley Davidson Enthusiast, May, June, July, &
August 1929. *Courtesy of the Doug Leikala Collection.* $20-$75 each

Four Copies Of The Harley Davidson Enthusiast, Sept., Oct., Nov., &
Dec., 1929. *Courtesy of the Doug Leikala Collection.* $20-$75 each

Harley Davidson Enthusiasts, Aug. 1931, June 1932, & December 1932. *Courtesy of the Dunbar Moonlighter Collection.* $15-$50 each

Harley Davidson Enthusiasts, January, February, April, May, & June 1933. *Courtesy of the Dunbar Moonlighter Collection.* $15-$50 each

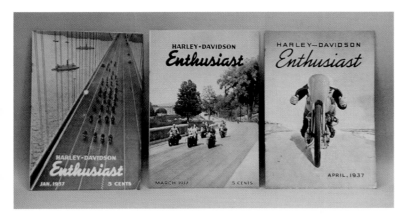

Harley Davidson Enthusiasts, January, March, & April 1937. *Courtesy of the Dunbar Moonlighter Collection.* $15-$50 each

December 1936 Harley Davidson Enthusiast, Cover: Santa with toy bag on a new Harley, features Christmas gifts. *Courtesy of the Dunbar Moonlighter Collection.* $15-$50 each

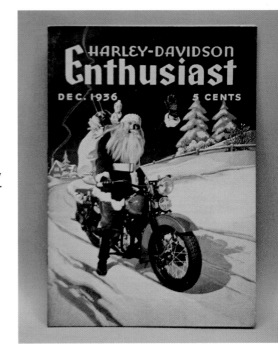

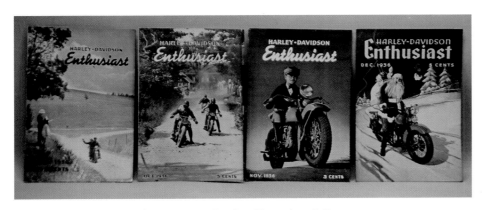

Harley Davidson Enthusiasts, August, October, November, & December 1936. *Courtesy of the Dunbar Moonlighter Collection.* $15-$50 each

Harley Davidson Enthusiasts, May, June, July, & September 1937.
Courtesy of the Dunbar Moonlighter Collection. $15-$50 each

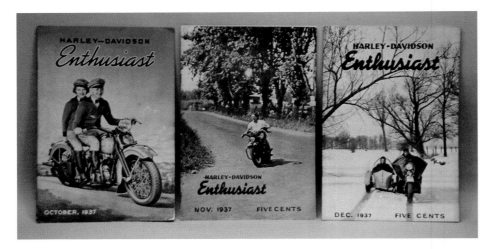

Harley Davidson Enthusiasts, October, November, & December 1937.
Courtesy of the Dunbar Moonlighter Collection. $15-$50 each

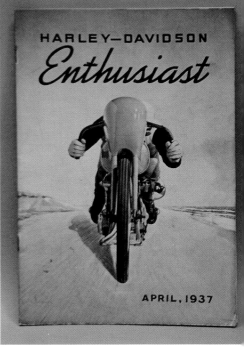

April 1937 Harley Davidson Enthusiast, cover: action portrayal Daytona
Speed Record on "61" OHV by Joe Patrali. *Courtesy of the Dunbar
Moonlighter Collection.* $15-$50

Harley Davidson Enthusiasts, January, February, & March 1938. *Courtesy of the Dunbar Moonlighter Collection.* $15-$50 each

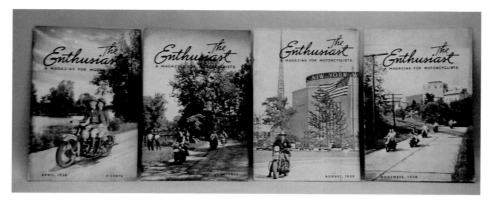

Harley Davidson Enthusiasts, April, June, August, & November 1938.
Courtesy of the Dunbar Moonlighter Collection. $15-$50 each

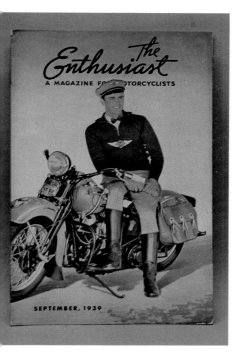

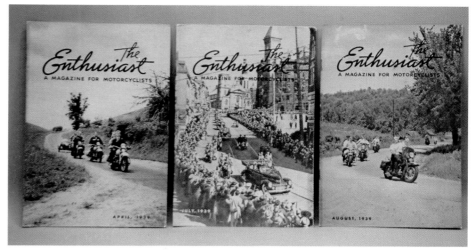

Harley Davidson Enthusiasts, April, July, & August 1939. *Courtesy of the Dunbar Moonlighter Collection.* $15-$50 each

Sept. 1939 Harley Davidson Enthusiast, cover: new 1940 Harley with rider in Harley garb, features photos all models. *Courtesy of the Dunbar Moonlighter Collection.* $15-$50

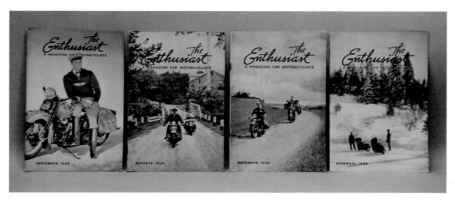

Harley Davidson Enthusiasts, September, October, November, & December 1939. *Courtesy of the Dunbar Moonlighter Collection.* $15-$50 each

Harley Davidson Enthusiasts, January-June 1940. *Courtesy of the Dunbar Moonlighter Collection.* $15-$50 each

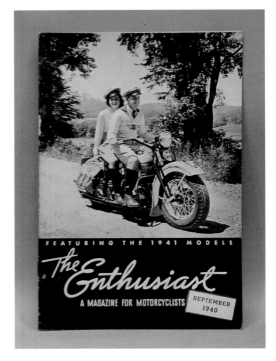

September 1940 Harley Davidson Enthusiast, cover: new Harley, full dressed couple, featuring new 1941 models. *Courtesy of the Dunbar Moonlighter Collection.* $15-$50

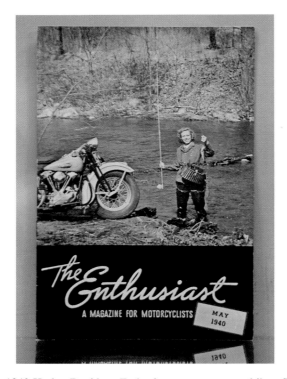

May 1940 Harley Davidson Enthusiast, cover: woman biker, fishing in Pascac River. *Courtesy of the Dunbar Moonlighter Collection.* $15-$50

Harley Davidson Enthusiasts, July-December 1942. *Courtesy of the Dunbar Moonlighter Collection.* $15-$50 each

Harley Davidson Enthusiasts, January, March, April, May, & June 1941.
Courtesy of the Dunbar Moonlighter Collection. $15-$50 each

Harley Davidson Enthusiasts, July-December 1941. *Courtesy of the Dunbar Moonlighter Collection.* $15-$50 each

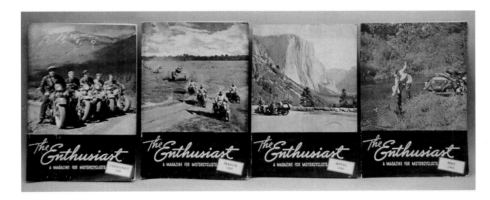

Harley Davidson Enthusiasts, February-May 1942. *Courtesy of the Dunbar Moonlighter Collection.* $15-$50 each

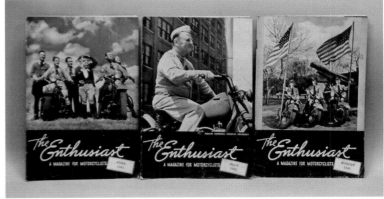

Harley Davidson Enthusiasts, June-August 1942. *Courtesy of the Dunbar Moonlighter Collection.* $15-$50 each

Harley Davidson Enthusiasts, September-December 1942. *Courtesy of the Dunbar Moonlighter Collection.* $15-$50 each

September 1942 Harley Davidson Enthusiast, cover: Clark Gable on dressed out Harley, Gene Autry in centerfold. *Courtesy of the Dunbar Moonlighter Collection.* $15-$50 each

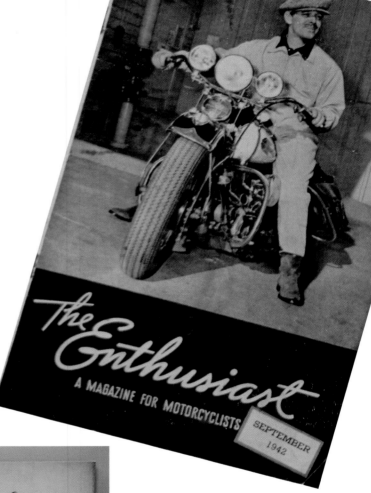

Harley Davidson Enthusiasts, January, March, & April 1943. *Courtesy of the Dunbar Moonlighter Collection.* $15-$50 each

Harley Davidson Enthusiasts, front and back covers for May, July, &
August 1943, back covers: Robert Young on his full dress Harley, Gene
Tierney on John Paynes' Harley, Van Johnson & Keenan Wynn both
on Harleys. *Courtesy of the Dunbar Moonlighter Collection.* $15-$50
each

July 1943 Harley Davidson Enthusiast, back cover: Gene Tierney on
John Paynes' Harley. *Courtesy of the Dunbar Moonlighter Collection.* $15-$50

August 1943 Harley Davidson Enthusiast, back cover: Van Johnson &
Keenan Wynn both on Harleys. *Courtesy of the Dunbar Moonlighter
Collection.* $15-$50

October 1943 Harley Davidson Enthusiast, back cover: Barbara Stanwyck with Bob Taylor and his "74" OHV Harley. *Courtesy of the Dunbar Moonlighter Collection.* $15-$50

Harley Davidson Enthusiasts, February, March, May, & August 1944. Back covers: Andy Devine & Dennis Morgan on Andy's two Harleys. *Courtesy of the Dunbar Moonlighter Collection.* $15-$50 each

Harley Davidson Enthusiasts, September-November 1943. *Courtesy of the Dunbar Moonlighter Collection.* $15-$50 each

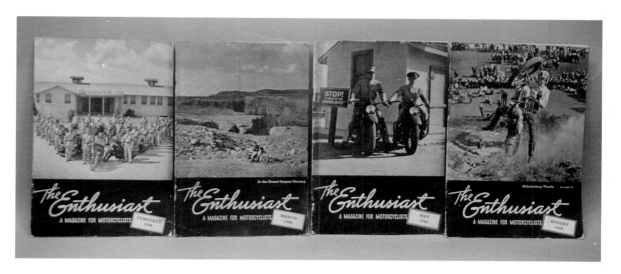

Harley Davidson Enthusiasts, February, March, May, & August 1944. *Courtesy of the Dunbar Moonlighter Collection.* $15-$50 each

Harley Davidson Enthusiasts, February, March, May, & August 1944. Back covers: Peggy Maley & Bill Shawn on Bill's "61" OHV Harley, Hollywood Get Together At Fleming Home, group of celebrities posed with four (4) Harleys, Carol Landis posed sidesaddle on Harley, Clark Gable, Ward Bond, Victor Fleming posed on Harleys. *Courtesy of the Dunbar Moonlighter Collection.* $15-$50 each

A closer look at two back covers, Peggy Maley & Bill Shawn and Clark Gable, Ward Bond, and Victor Fleming, all posed on their Harleys. *Courtesy of the Dunbar Moonlighter Collection.* $15-$50 each

Harley Davidson Enthusiasts, January-April 1944. *Courtesy of the Dunbar Moonlighter Collection.* $15-$50 each

Harley Davidson Enthusiasts, May-August 1944. *Courtesy of the Dunbar Moonlighter Collection.* $15-$50 each

Harley Davidson Enthusiast, August 1944. *Courtesy of the Dunbar Moonlighter Collection.* $15-$50 each

Harley Davidson Enthusiasts, September-December 1944. *Courtesy of the Dunbar Moonlighter Collection.* $15-$50 each

Harley Davidson Enthusiasts, January-March 1945. *Courtesy of the Dunbar Moonlighter Collection.* $15-$50 each

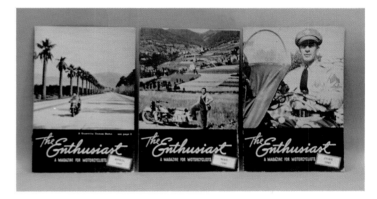

Harley Davidson Enthusiasts, April-June 1945. *Courtesy of the Dunbar Moonlighter Collection.* $15-$50 each

Harley Davidson Enthusiasts, July-September 1945. *Courtesy of the Dunbar Moonlighter Collection.* $15-$50 each

Harley Davidson Enthusiasts, October-December 1945. *Courtesy of the Dunbar Moonlighter Collection.* $15-$50 each

Harley Davidson Enthusiasts, January-April 1946. *Courtesy of the Dunbar Moonlighter Collection.* $10-$35 each

Harley Davidson Enthusiasts, September-December 1946. *Courtesy of the Dunbar Moonlighter Collection.* $10-$35 each

Harley Davidson Enthusiasts, May, June, & August 1946. *Courtesy of the Dunbar Moonlighter Collection.* $10-$35 each

Harley Davidson Enthusiasts, January-April 1947. *Courtesy of the Dunbar Moonlighter Collection.* $10-$35 each

Harley Davidson Enthusiasts, May-August 1947. *Courtesy of the Dunbar Moonlighter Collection.* $10-$35 each

Harley Davidson Enthusiasts, September-December 1947. *Courtesy of the Dunbar Moonlighter Collection.* $10-$35 each

Harley Davidson Enthusiasts, January-April 1948. *Courtesy of the Dunbar Moonlighter Collection.* $10-$35 each

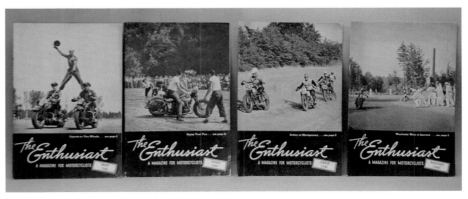

Harley Davidson Enthusiasts, May-August 1948. *Courtesy of the Dunbar Moonlighter Collection.* $10-$35 each

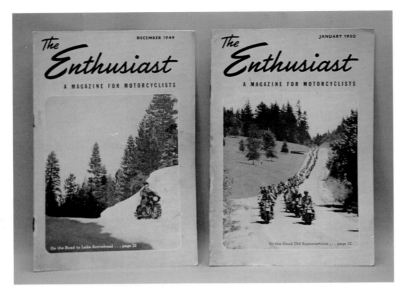

Harley Davidson Enthusiasts, December 1949, January 1950. *Courtesy of the Dunbar Moonlighter Collection.* $10-$35 each

Harley Davidson Enthusiasts, January, March, May, & June 1951. *Courtesy of the Dunbar Moonlighter Collection.* $10-$35 each

Harley Davidson Enthusiasts, February, June, July, & September 1952. *Courtesy of the Dunbar Moonlighter Collection.* $10-$35 each

May 1956 Harley Davidson Enthusiast With The King, Elvis Presley. Harley Davidson recently purchased this exact motorcycle. One of the most sought after copies of the Enthusiast. *Courtesy of Ken Kalustian.* $50-$200

Volume I, Issue I of The American Motorcycling Association First Magazine, January 1947. *Courtesy of Ken Kalustian.* $30-$85

Indian News Magazines, January/February, April/May, & June 1928. *Courtesy of the Dunbar Moonlighter Collection.* $20-$75 each

Indian News Magazines, September/October & November/December 1928. *Courtesy of the Dunbar Moonlighter Collection.* $20-$75 each

Indian News Magazines, February, March/April, & May 1929. *Courtesy of the Dunbar Moonlighter Collection.* $20-$75 each

Indian News Magazines, June/July, August, September/October, & Special, 1929. *Courtesy of the Dunbar Moonlighter Collection.* $20-$75 each

Indian News Magazines, March-May 1930. *Courtesy of the Dunbar Moonlighter Collection.* $20-$75 each

Indian News Magazines, June, July, & August 1930. *Courtesy of the Dunbar Moonlighter Collection.* $20-$75 each

Indian News Magazines, September-December 1930. *Courtesy of the Dunbar Moonlighter Collection.* $20-$75 each

Indian News Magazines, March, June, & July/August 1931. *Courtesy of the Dunbar Moonlighter Collection.* $20-$75 each

Indian News Magazines, September/October & November/December 1931. *Courtesy of the Dunbar Moonlighter Collection.* $20-$75 each

Indian News Magazine, June/July 1932. *Courtesy of the Dunbar Moonlighter Collection.* $20-$75 each

Indian News Magazine, November/December 1932. *Courtesy of the Dunbar Moonlighter Collection.* $20-$75 each

Indian News Magazines, January/February, June/July, & November/December 1932. *Courtesy of the Dunbar Moonlighter Collection.* $20-$75 each

Copies Of The Indian Wigwam News — June 1943, December 1943, January/February 1944, & March/April 1944, 23 pages, featuring Indian efforts in WWI & WWII. *Courtesy of the Dunbar Moonlighter Collection.* $10-$65 each

Indian Instruction Books, 1910-1920. *Courtesy of Cris and Pat Simmons.* $50-$250 each

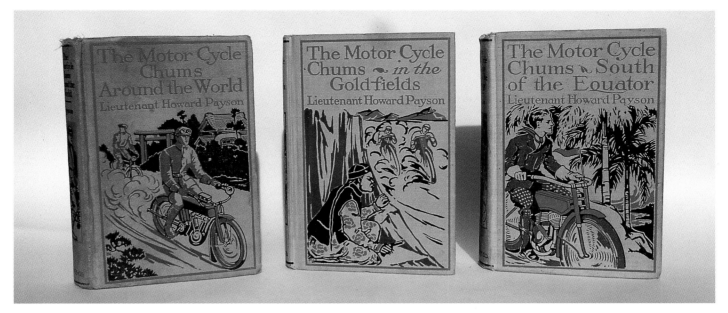

Motorcycle Chums Around The World, Motorcycle Chums In The Goldfields, The Motorcycle Chums South Of The Equator, by Lieutenant Howard Payson, ca. 1920. *Courtesy of Cris & Pat Simmons.* $20-$60 each

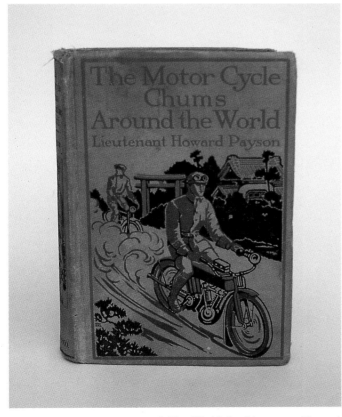

The Motorcycle Chums Around The World, by Lieutenant Howard Payson, ca. 1920. *Courtesy of Cris & Pat Simmons.* $20-$60

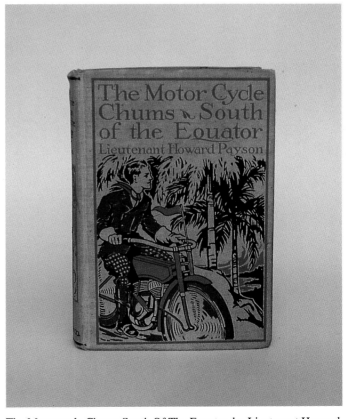

The Motorcycle Chums South Of The Equator, by Lieutenant Howard Payson, ca. 1920. *Courtesy of Cris & Pat Simmons.* $20-$60

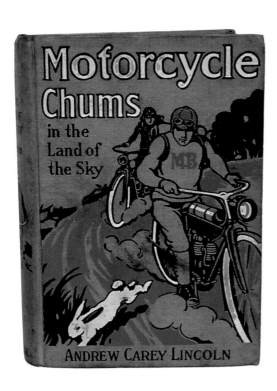

The Motorcycle Chums In The Land Of The Big Sky, by Lieutenant Howard Payson, ca. 1920. *Courtesy of Cris & Pat Simmons.* $20-$60

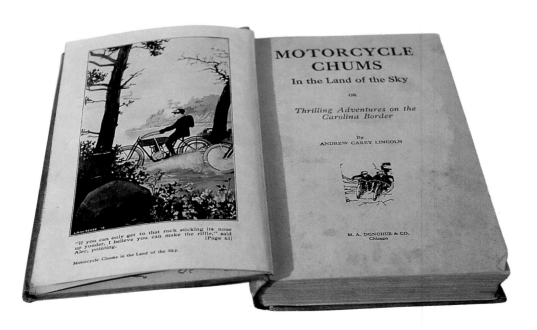

The Motorcycle Chums In The Land Of The Big Sky, by Lieutenant Howard Payson, ca. 1920. *Courtesy of Cris & Pat Simmons.* $20-$60

Five Motorcycle Chums Books, by Lieutenant Howard Payson, ca. 1920. *Courtesy of Cris & Pat Simmons.* $20-$60 each

The Big Five Motorcycle Boys Books, by Ralph Marlow, ca. 1920. *Courtesy of Cris & Pat Simmons.* $20-$60 each

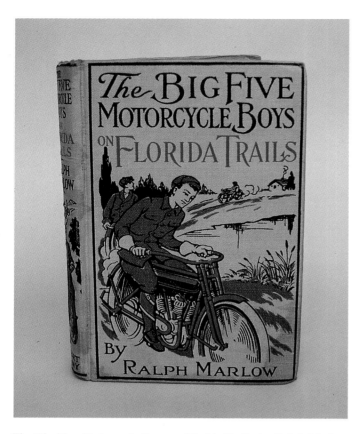

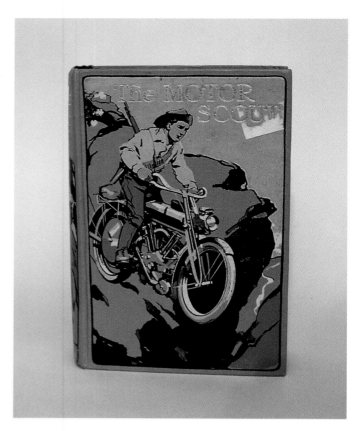

The Big Five Motorcycle Boys — Florida Trails, by Ralph Marlow, ca. 1920. *Courtesy of Cris & Pat Simmons.* $20-$60

The Motor Scout, ca. 1920. *Courtesy of Cris & Pat Simmons.* $20-$60

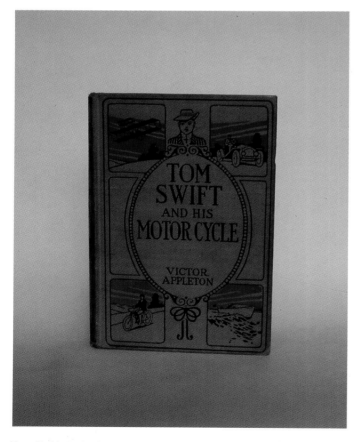

The Motor Scout, ca. 1920. *Courtesy of Cris & Pat Simmons.* $20-$60

Tom Swift And His Motorcycle, by Victor Appleton, ca. 1920. *Courtesy of Cris & Pat Simmons.* $20-$60

FAM & AMA AWARDS

FAM Ribbons For Annual Meets, 1904, 1905, Cambridge, Maryland,
July 8, 9, 1904, Waltham, Massachusetts Aug. 8-12, 1905 with FAM
badge. *Courtesy of the Doug Leikala Collection.* $200-$400 each

CHAPTER IV

THE FAM or Federation of American Motorcyclists, created between 1901-1904, was the first national governing body of the fresh new cyclists who were whizzing about the countryside. Always strapped for cash and constantly under attack for uneven racing standards and practices, and for weak leadership, the FAM gave way to the AMA in 1922. The AMA still exists as the arm of the motorcycling world. Because the FAM only existed for a relatively short amount of time, any pins, fobs, signs, or related FAM items are highly collectible. Amazingly, thanks to the efforts of longtime collector Doug Leikala, there are a number of ultra rare FAM items photographed in this book.

To foster good fellowship and reward membership, the AMA gave out annual Gypsy Tour awards to members (and still do even to this day) from belt buckles to fobs, to bracelets, to tie clips. How many bikers actually wear tie clips? For a number of years, women were given separate awards, which are scarcer and therefore more valuable. Likewise, any fobs that are engraved for winning an FAM or AMA racing competition are very collectible, the older the better, the higher the place the better, and the bigger the name in racing the better.

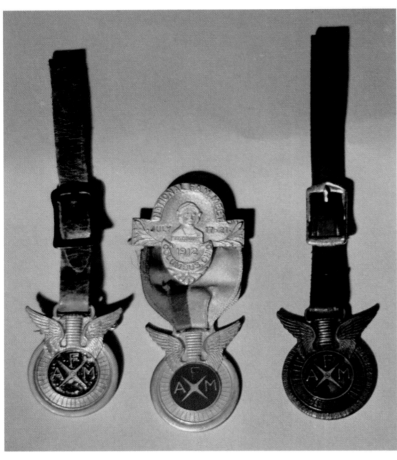

FAM Fobs, 1909, 1912, 1914. *Courtesy of the Doug Leikala Collection.* $300-$600 each

FAM Fobs, ca. 1915. *Courtesy of the Doug Leikala Collection.* $200-$350 each

FAM Motorcycle Races Program, July 15, 1911, Fort Erie. *Courtesy of the Doug Leikala Collection.* $100-$350

Ninth Annual FAM Meet Program, Buffalo, New York, FAM Helping Handbook, ca. 1912. *Courtesy of the Doug Leikala Collection.* $100-$300 each

Milwaukee Journal Tour Watch Fobs, 1912, 1913, 1914. *Courtesy of the Doug Leikala Collection.* $200-$350 each

A 1915 Endurance Run Award, Invincible Motorcycle Club, ca. 1915. $200-$400. Racing Award, ca. 1950s. *Courtesy of the Doug Leikala Collection.* $75-$150

A 1916 FAM Pin, gold. *Courtesy of the Dunbar Moonlighter Collection.* $100-$300

Milwaukee Journal Tour Watch Fobs, 1915 & 1916. *Courtesy of the Doug Leikala Collection.* $200-$350 each

Motorcycle Club Award, ca. 1908, $300-$600; the 1916 Endurance Run, Invincible Motorcycle Club. *Courtesy of the Doug Leikala Collection.* $300-$600

PMC Triangle Run Award Medals, 1916, 1920, 1921. *Courtesy of the Doug Leikala Collection.* $200-$400

Racing Fob & Harley With Sidecar Fob, both ca. 1915. *Courtesy of the Doug Leikala Collection.* $200-$350

A 1917 Mount Beacon Hillclimb Award. $200-$400. Racing Award, ca. 1940s. *Courtesy of the Doug Leikala Collection.* $75-$150

Reading Motorcycle Club Fob, Crotona Motorcycle Club 24 Hour Endurance Run Award, Racing Fob, ca. 1915. *Courtesy of the Doug Leikala Collection.* $200-$450 each

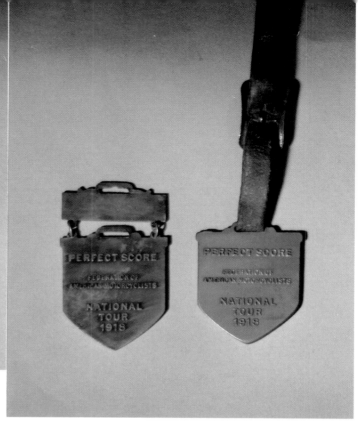

A pair of FAM 1918 Gypsy Tour Awards, Perfect Score Award, for FAM member & non-member. *Courtesy of the Doug Leikala Collection.* Member $150-$350, Non-member $200-$350

A pair of FAM Gypsy Tour Awards, a 1917 Perfect Score Award Fob, $150-$350. Three Year Fob, 1917-1919. *Both courtesy of the Doug Leikala Collection.* $300-$550.

Three Tire Awards Commemorating 1920 Gypsy Tour, 1919 National Championship and 1917 NE Gypsy Tour. *Courtesy of the Doug Leikala Collection.* $300-$600

A pair of FAM/M. and A.T.A. 1919 Gypsy Tour Awards, Perfect Score Award, Leather Fob & Ribboned Pin, for FAM member & non-member. *Courtesy of the Doug Leikala Collection.* Member $150-$350, Non-member $200-$350

AMA 1920 National Motorcycle Gypsy Tour Men's Award Leather Fob. *Courtesy of the Doug Leikala Collection.* $150-$300

A pair of AMA 1922 Gypsy Tour Awards, Perfect Score Award, Men's Leather Fob, & Ladies' Pinback. *Courtesy of the Doug Leikala Collection.* Men's $150-$300, Ladies' $200-$350

AMA 1921 Gypsy Tour Men's Leather Fob Award. *Courtesy of the Doug Leikala Collection.* $150-$300

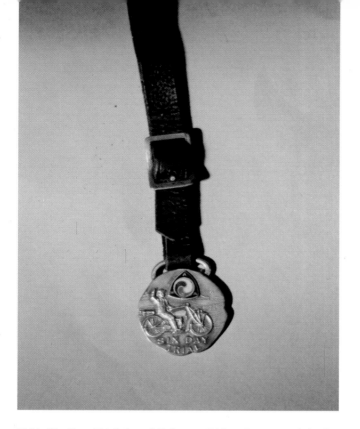

AMA Six Day Trial Award Fob, ca. 1920s. *Courtesy of the Doug Leikala Collection.* $200-$600

A pair of 1923 AMA/M. and A.T.A. Gypsy Tour Awards, Men's Leather Fob, Ladies' Pinback. *Courtesy of the Doug Leikala Collection.* Men's $150-$300, Ladies' $200-$350

A pair of 1924 AMA Gypsy Tour Awards, 1924, Perfect Score, Men's Leather Fob, Ladies' Pinback. *Courtesy of the Doug Leikala Collection.* Men's $150-$300, Ladies' $200-$350

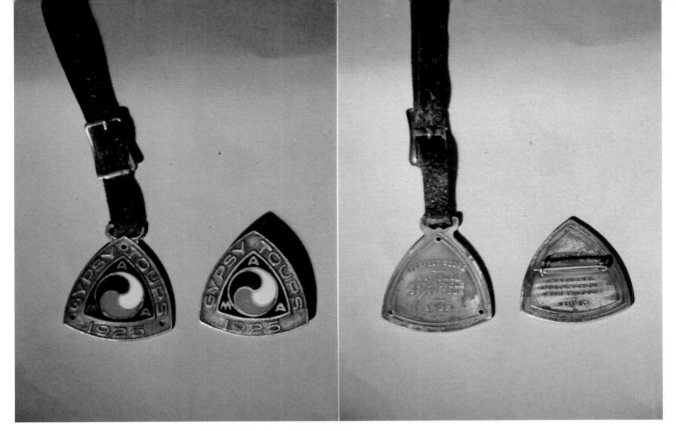

A pair of 1925 AMA Gyspy Tour Awards Men's Leather Fob Ladies'
Pinback. *Courtesy of the Doug Leikala Collection.* Men's $150-$300,
Women's $200-$350

A pair of 1926 National Motorcycle Gypsy Tour Awards, Men's Leather
Fob, Ladies' Ribbon. *Courtesy of the Doug Leikala Collection.* Mens'
$150-$300, Ladies' $200-$350

Philadelphia Motorcycle Association, ca. 1920s. $200-$400. Two Racing Award Fobs, ca. 1950s. *Courtesy of the Doug Leikala Collection.* $75-$150

Crotona Motorcycle Club Award, ca. 1915, $200-$450. A 1930 200 Mile National Championship TT Award, Keene, New Hampshire, $100-$200. Long Distance Tourist Award, ca. 1920s. *Courtesy of the Doug Leikala Collection.* $200-$400

AMA National Competition Medal, Second Place, 1933. *Courtesy of the Doug Leikala Collection.* $200-$400

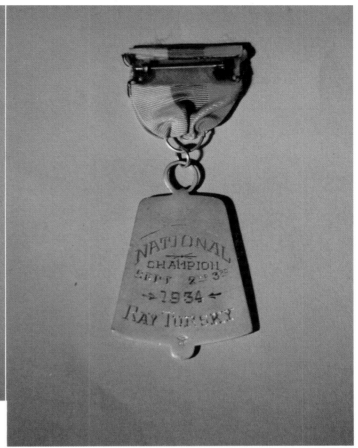

A 1934 Jack Pine 1st Prize Medal, National Champion, Ray Tursky, American Motorcycle Association. *Courtesy of the Doug Leikala Collection.* $300-$700

A 1935 Jack Pine Second Prize Medal, AMA, Class A Solo, Ray Tursky. *Courtesy of the Doug Leikala Collection.* $300-$700

  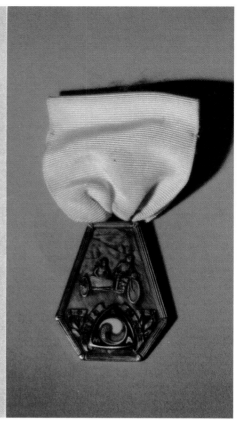

AMA Second Place Silver Medal, ca. 1930s. *Courtesy of Bob "Sprocket" Eckardt.* $100-$300

AMA Third Place Bronze Medal, ca. 1930s. *Courtesy of Bob "Sprocket" Eckardt.* $100-$300

AMA Third Place Medal, ca. 1930s. *Courtesy of the Doug Leikala Collection.* $100-$300

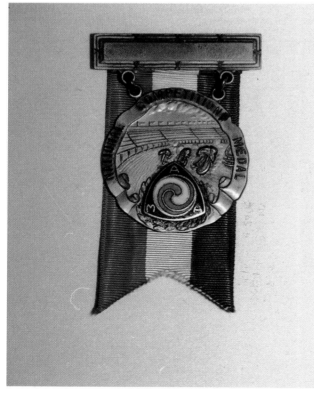 

A 1938 AMA National Championship Medal, TT Races, Laconia, New Hampshire. This award was won by Waikko "Farmer" Karjalainen. *Courtesy of Bob "Sprocket" Eckardt.* $250-$500

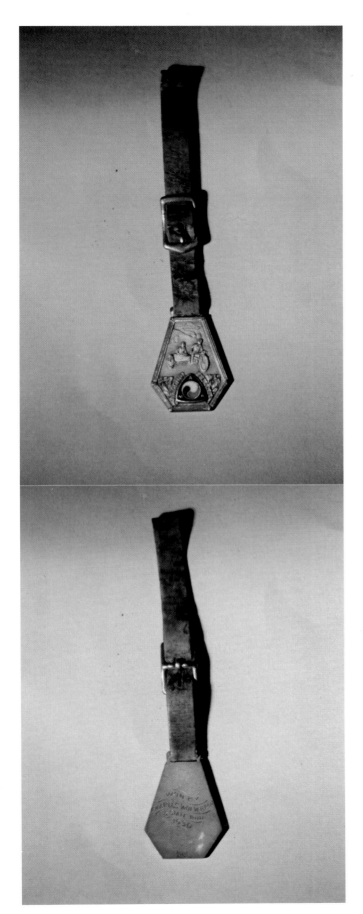

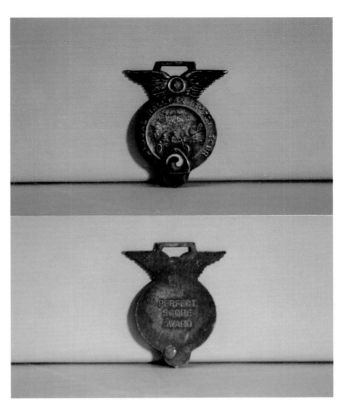

A 1930 AMA Gypsy Tour Perfect Score Award, Annual Rally, with AMA logo, embossed cycle & sidecar, 1 1/2" W x 2" H. *Courtesy of the Dunbar Moonlighter Collection.* $100-$250

A 1930 Second Place AMA Award Fob. *Courtesy of the Doug Leikala Collection.* $100-$300

A 1940 AMA Merit Award Fob. *Courtesy of the Doug Leikala Collection.* $100-$300

A 1932 AMA Gypsy Tour Award Ring. *Courtesy of the Doug Leikala Collection.* $200-$500

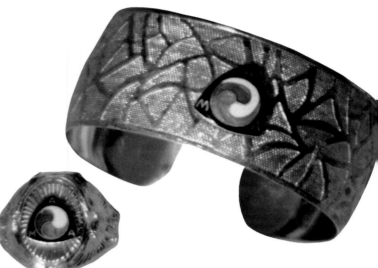

AMA Gypsy Tour Men's & Ladies' Awards for 1936, ring & bracelet. *Courtesy of the Doug Leikala Collection.* $100-$300, $300-$750

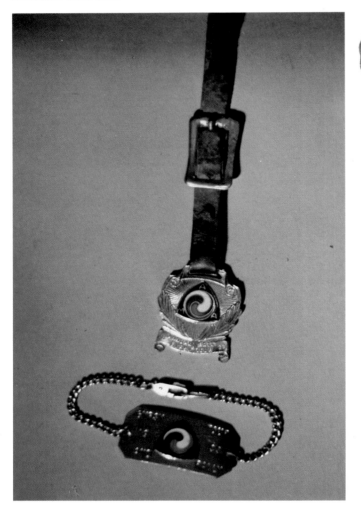

AMA Gypsy Tour Men's & Ladies' Awards for 1933, fob & bracelet. *Courtesy of the Doug Leikala Collection.* $100-$300, $200-$500

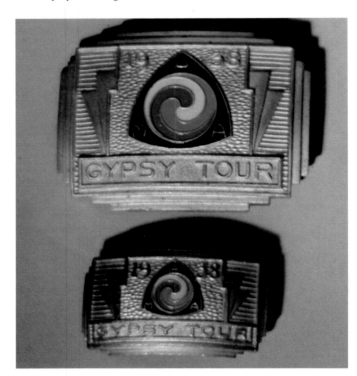

AMA Gyspy Tour Men's & Ladies' Award Belt Buckles for 1938. *Courtesy of the Doug Leikala Collection.* $200-$350 each

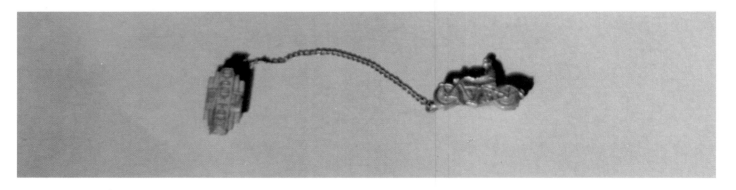

A 1937 AMA Gypsy Tour Ladies Chain Pin. *Courtesy of the Doug Leikala Collection.* $150-$350

A 1941 AMA Gypsy Tour Award Badge. *Courtesy of the Doug Leikala Collection.* $50-$175

AMA Gypsy Tour Men's & Ladies' Awards for 1939, medal & bracelet. *Courtesy of the Doug Leikala Collection.* $100-$250 each

AMA Gyspy Tour Men's & Ladies' Award of 1940, key chain & bracelet. *Courtesy of the Doug Leikala Collection.* $200-$450, $100-$250

AMA Gypsy Tour Men's & Ladies' Awards of 1942, belt buckle & brooch. *Courtesy of the Doug Leikala Collection.* $100-$300, $100-$250

A pair of 1942 and 1943 AMA Patriotic Award Plaques. During WWII, these plaques were given to clubs in lieu of the normal award jewelry as part of the war effort. *Courtesy of the Doug Leikala Collection.* $75-$150 each

A 1948 AMA Gypsy Tour Award Pin. *Courtesy of the Doug Leikala Collection.* $50-$125

AMA Gypsy Tour Award Tie Clip of 1949, AMA Ladies' Gypsy Tour Award Bracelet. *Courtesy of the Doug Leikala Collection.* $50-$150 each

The 1944 and 1945 AMA Patriotic Contest Plaques. During WWII, these plaques were given to clubs in lieu of the normal award jewelry as part of the war effort. *Courtesy of the Doug Leikala Collection.* $75-$150 each

AMA Gypsy Tour Men's & Women's Award Rings of 1946. *Courtesy of the Doug Leikala Collection.* $100-$250 each

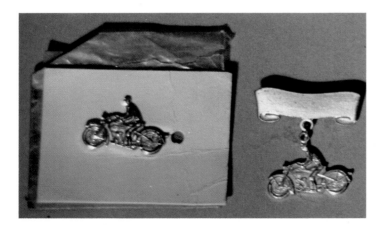

AMA Gyspy Tour Awards, Ladies' Pin, Men's Medal of 1947. *Courtesy of the Doug Leikala Collection.* $75-$200 Each

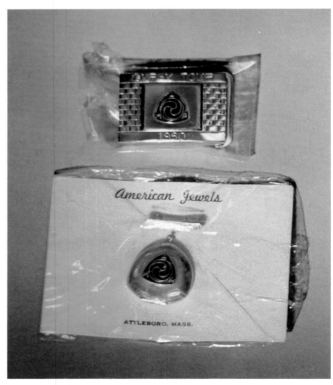

A 1950 AMA Gypsy Tour Award Belt Buckle; 1950 AMA Ladies' Gypsy Tour Pin, American Jewels, Attleboro, Massachusetts. *Courtesy of the Doug Leikala Collection.* $50-$125 each

A 1951 AMA Gyspy Tour Award Keychain; 1951 Ladies' Gypsy Tour Award Bracelet. *Courtesy of the Doug Leikala Collection.* $50-$125 each

A 1952 AMA Gypsy Tour Award Glass Ashtray. *Courtesy of the Doug Leikala Collection.* $25-$75

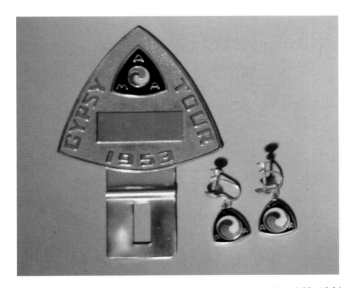

AMA Gypsy Tour Award License Plate Mount of 1953, 1953 AMA Gypsy Tour Ladies' Earrings. *Courtesy of the Doug Leikala Collection.* $50-$125 each

A 1954 AMA Gypsy Tour Winged Pin Award. *Courtesy of the Doug Leikala Collection.* $50-$125

A 1955 AMA Gypsy Tour Award Belt Buckle. *Courtesy of the Doug Leikala Collection.* $50-$125

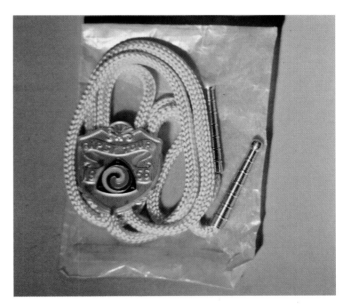

A 1956 AMA Lariat Gypsy Tour Award, with cord & package. *Courtesy of the Doug Leikala Collection.* $50-$125

A 1957 Gypsy Tour Award Badge. *Courtesy of the Doug Leikala Collection.* $50-$125

AMA Award Belt Buckle of 1958, AMA Logo, 2 1/2" W x 1 3/4" H. 1958 AMA Gypsy Tour Award Badge. *Courtesy of the Dunbar Moonlighter Collection.* $50-$125.

A 1969 AMA Sportsman Champion Pin. *Courtesy of the Doug Leikala Collection.* $50-$75

A 1961 AMA Award Tie Clip, Embossed AMA Logo, 2 1/4" W x 1" H, on original card. $50-$125. 1963 AMA Gypsy Tour Award Tie Clasp, 1 3/4" L, on original card. *Courtesy of the Dunbar Moonlighter Collection.* $50-$125

A 1973 AMA Safety Award Pin. *Courtesy of the Doug Leikala Collection.* $50-$75

AMA Gypsy Tour Award Pin, 1954; a Harley Davidson Twin Engine Pin, a 1962 AMA Award Pin, AMA logo, on original card. *Courtesy of the Dunbar Moonlighter Collection.*

Dayton Ramblers Club Pin, ca. 1950s. *Courtesy of the Doug Leikala Collection.* $25-$65

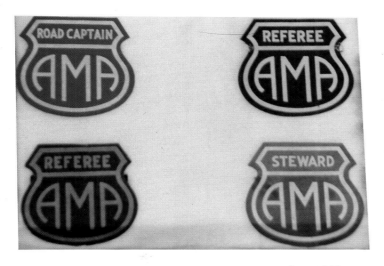

Set Four AMA Patches - Road Captain, Referee, Steward, ca. 1940s-1950s. *Courtesy of the Doug Leikala Collection.* $25-$75 each

AMA Patches. *Courtesy of the Doug Leikala Collection.*

AMA Tour Award Patches — 1964, 1965, 1966, & 1967. *Courtesy of the Doug Leikala Collection.* $10-$40 each

Fifty years of AMA Tour award patches, 1938-1988. *Courtesy of the Doug Leikala Collection.* $10-$85 each.

AMA Award, ROG Belt Buckle with AMA Logo, 1940s-1980s. *Courtesy of the Doug Leikala Collection.* $30-$125 each

The 1980s & 1990s AMA Gypsy Tour Award Badges & Fob. *Courtesy of the Doug Leikala Collection.* $5-$25 each

AMA Ceramic Gypsy Tour Mug, ca. 1940s, Ceramic Plate, Petersen Bros. Motorcycles, Hayward, California, ca. 1920s. *Courtesy of the Doug Leikala Collection.* $50-$150, $100-$350

DEALER JEWELRY
& PROMOS

Excelsior Auto Cycle Honing Stone, ca. 1915. *Courtesy of Bob "Sprocket" Eckardt.* $100-$600

Is there such a thing as something for nothing? Well maybe, when you consider the rich legacy of giveaways and free samples continued even to this day. Look around your house and count up the number of pencils, pens, screwdrivers, pins, buttons, note pads, magnets, and rain hats (always from insurance companies, what are they trying to say?) from various outfits stumping for your business. Every year in February on In-

dian Day, the day where the company would unveil the newest Indian models, prospective customers would be feted with giveaway winged pins, etc. Since pocket watches were popular in the early 1900s, fobs were commonly handed out, bearing the proud X logo of Excelsior, or the figural arrowhead relic of Indian. Today collectors scour glass cases in search of "smalls" to fill in their own collections.

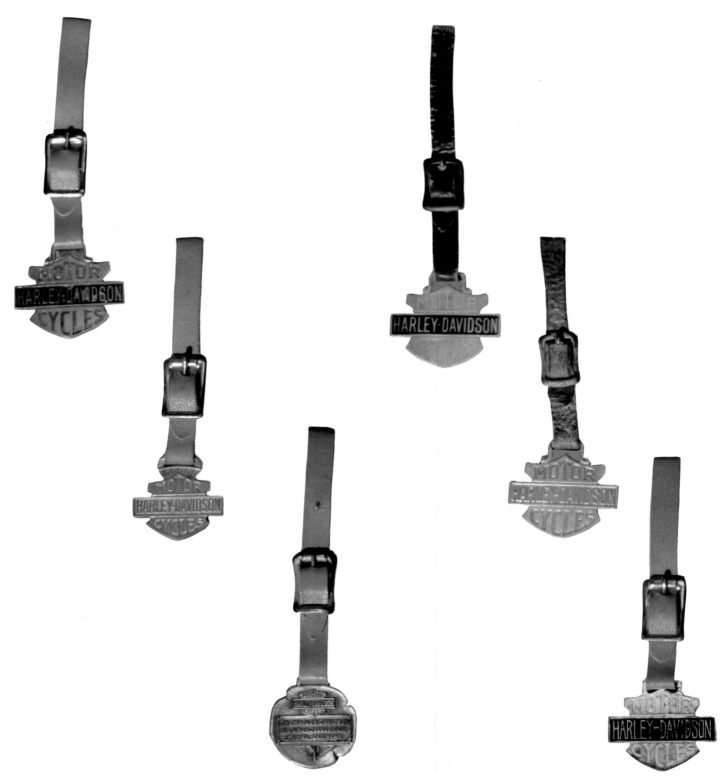

Three Harley Fobs, ca.1915-1920s. *Courtesy of the Doug Leikala Collection.* $200-$350 each

Three Eras Of Harley Davidson Fobs, ca. 1915, 1920s, & 1930s. *Courtesy of the Doug Leikala Collection.* $200-$350 each.

Pair Racing Fobs, one Angsten Koch Co., Chicago Motorcycle Specialties Co., ca. 1915. *Courtesy of the Doug Leikala Collection.* $200-$400 each

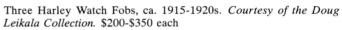

Three Harley Watch Fobs, ca. 1915-1920s. *Courtesy of the Doug Leikala Collection.* $200-$350 each

Excelsior Watch Fobs, ca. 1920s. *Courtesy of the Doug Leikala Collection.* $100-$350 each

Four Indian Motocycle Fobs, ca. 1915-1925. *Courtesy of the Doug Leikala Collection.* $150-$350 each

Indian Figural Headdress Fob, cloisonne, ca. 1930s. *Courtesy of the Doug Leikala Collection.* $200-$400

Four Indian Figural Headdress Fobs, ca. 1920s. *Courtesy of the Doug Leikala Collection.* Each $150-$350

Indian Motorcycle Fobs, Indian logo in circle, ca. 1940s. $75-$200.
Arrowhead Promoting Indian 4, ca. 1930s. $100-$250. Figural Indian
Head, ca. 1920s. $150-$350. Indian Motocycles Arrowhead, ca. 1920s.
All courtesy of the Doug Leikala Collection. $100-$250

Excelsior Auto Cycle Fobs, ca. 1915-1925. *Courtesy of the Doug
Leikala Collection.* $150-$350

The 28th Semi Annual Convention, Washington State Peace Officers
Association Ribbon & Indian Arrowhead Fob, ca. 1920s. *Courtesy of
the Dunbar Moonlighter Collection.* $200-$400

Yale, Deluxe Motorcycle Club, Musselman Fobs, ca. 1915. *Courtesy of the Doug Leikala Collection.* $100-$350 each

Indian Cycle Miniature Embossed Pin, ca. 1920s, 1" L. *Courtesy of the Dunbar Moonlighter Collection.* $50-$100

Crazy Indian Motocycles Pin, ca. 1920s, Embossed "Indian Motorcycles, Eastern Bros., Rochester, NY," reverse side, 1" W x 1 1/2" H. $50-$100. Indian Motocycles Embossed Arrowhead Watch Fob, ca. 1920s, 1 3/4" H. $50-$200. Indian Motocycle Pin, ca. 1920s, silver wings gold Indian head, embossed Indian motocycle on wings, 2" Long. $100-$250 *Courtesy of the Dunbar Moonlighter Collection.*

Pair Harley Davidson Endurance, Power Speed Embossed Fobs, ca. 1930s. *Courtesy of the Doug Leikala Collection.* $150-$300 each

Group Of Fobs, featuring Yale, ca. 1915. *Courtesy of the Doug Leikala Collection.* Yale - $200-$350

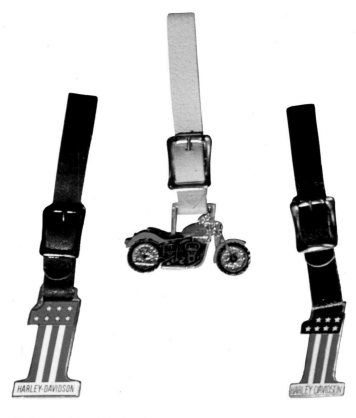

Harley Davidson Watch Fobs, ca. 1960s-1970s. *Courtesy of the Doug Leikala Collection.* $40-$85 each

Group of Fobs, featuring Ace 4 Fob, ca. 1920. *Courtesy of the Doug Leikala Collection.* Ace - $200-$350

Pair Harley Davidson Fobs, ca. 1950s. *Courtesy of the Doug Leikala Collection.* $50-$150 each

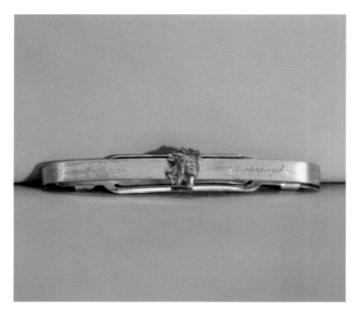

Indian Motorcycle Tie Clip, ca. 1930s, Indian Head, embossed "Indian Motorcycle" in red, 3" Long. *Courtesy of the Dunbar Moonlighter Collection.* $200-$350

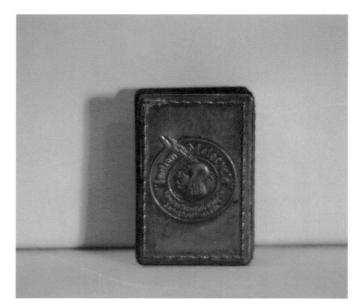

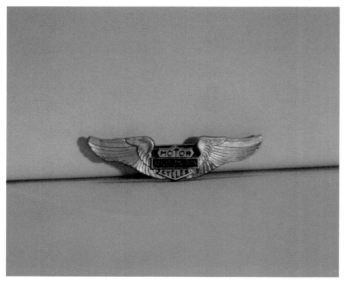

Harley Davidson Winged Pin with Logo, 2" L x 1/2" H. *Courtesy of the Dunbar Moonlighter Collection.* $50-$150

Indian Motocycles Matchsafe, circa 1915-1920, embossed Hendee Manufacturing Co., with Indian logo, on reverse embossed Highest Honors 1904 World's Fair in St. Louis, 1915 San Francisco Exposition. *Courtesy of the Dunbar Moonlighter Collection.* $75-$250

Trio Of Indian Matchsafes, ca. 1915. *Courtesy of the Doug Leikala Collection.* $75-$200 each

Indian Motorcycle Match Holder, 1920s, copper, embossed logo, 1 1/2" W x 2 1/4" H. *Courtesy of the Dunbar Moonlighter Collection.* $100-$300

Harley Davidson Twin Cylinder Pin, ca. 1930s, 1/2" W x 3/4" H, embossed. *Courtesy of the Dunbar Moonlighter Collection.* $50-$125

Harley Davidson Diecut Bar Spinner & Bottle Opener, ca. 1930s, probably can pick the mud out of your boots, too. *Courtesy of the Doug Leikala Collection.* $100-$400

Harley Davidson Embossed Cigarette Case, ca. 1950s, Eagle logo, 2 1/2" W x 4" H. *Courtesy of the Dunbar Moonlighter Collection.* $200-$350

Harley Davidson Winged Pin, with logo, orange, gold & black, 3" Long. *Courtesy of the Dunbar Moonlighter Collection.* $25-$100

Harley Davidson Patch, ca. 1930s. *Courtesy of the Doug Leikala Collection.* $20-$50

Harley Davidson Silver Pin, winged logo, double stud backing, 2" Long. *Courtesy of the Dunbar Moonlighter Collection.* $50-$100

Harley Davidson Patch, winged logo, 10" Long. *Courtesy of the Dunbar Moonlighter Collection.* $50-$100

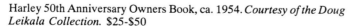

Harley 50th Anniversary Owners Book, ca. 1954. *Courtesy of the Doug Leikala Collection.* $25-$50

Harley Davidson Embossed Book, ca. 1920s. *Courtesy of the Doug Leikala Collection.* $50-$100

Indian Motocycles Leather Owner's Book, ca. 1920s. *Courtesy of the Doug Leikala Collection.* $50-$125

Harley Davidson Embossed Book, ca. 1920s. *Courtesy of the Doug Leikala Collection.* $50-$100

Brownie's Indian Motorcycles Book, ca. 1940s. *Courtesy of the Doug Leikala Collection.* $50-$100

Cable Harley Davidson Ashtray, ca. 1960s. *Courtesy of the Doug Leikala Collection.* $25-100

Harley Ashtray, ca. 1930s. $30-$100. Don Farrow's Harley Ashtray, ca. 1950s. $10-$50. *Courtesy of the Doug Leikala Collection.*

Pair Harley Tin Ashtrays, ca. 1950s. *Courtesy of the Doug Leikala Collection.* $25-$75 each

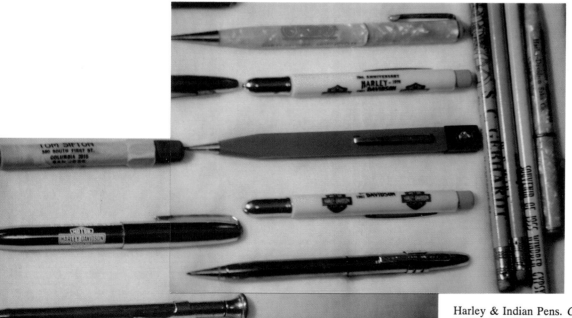

Harley & Indian Pens. *Courtesy of the Doug Leikala Collection.* $2-$10 each

Harley Pins & Badges, 1920s-1950s. *Courtesy of the Doug Leikala Collection.* $50-$150 Each

Harley Pins & Badges, 1920s-1980s. *Courtesy of the Doug Leikala Collection.* $50-$150 Each

Selection Of 6 Matchbook Covers, 1920s-1940s. *Courtesy of the Doug
Leikala Collection.* $5-$15 each

Selection Of 6 Matchbook Covers, 1920s-40s. *Courtesy of the Doug
Leikala Collection.* $5-$15 each

POSTCARDS & PHOTOGRAPHS

Racer Russell Smiley Goodyear Postcard, ca. 1915. *Courtesy of the Doug Leikala Collection.* $50-$100

Smile! From all the photos one can find in biker publications, it seems that bikers enjoy having themselves photographed for posterity, er, let me rephrase that. Early postcards, some from dealers, some home-made, show bikers happily rolling along, or posing with that satisfied "this is where I want to be" look. Or, collectors can look for photographs of motorcycle clubs and gypsy tour gatherings, with cycles lined up like dominoes, stretching out as far as the lens would allow. Occasionally photographs will surface of bikers in front of their home town dealers, great bits of history showing the look or "eye" of the time.

A Pair Of Curtiss Motorcycle Postcards, including World Beach Record by Glenn Curtiss, ca. 1910. *Courtesy of the Doug Leikala Collection.* $25-$100 each

Firestone Motorcycle Tires Advertising Postcard, ca. 1920s. *Courtesy of the Doug Leikala Collection.* $30-$60

A Pair Of Indian Motocycle Envelopes, ca. 1910. *Courtesy of the Doug Leikala Collection.* $50-$150 each

Postcards — Columbus Motordrome, Columbus Ohio, Record Breaking Sears Motorcycle, Motordrome At Luna Park, Coney Island, ca. 1915-1920s. *Courtesy of the Doug Leikala Collection.* $25-$100 each

Wagner Dealership Postcard, ca. 1915. Harley Postcard, ca. 1910.
Courtesy of the Doug Leikala Collection. $20-$30 each

Motorcycle Postcards, Emblem, Flanders, Indian Race, ca. 1915. *Courtesy of the Doug Leikala Collection.* $25-$100 each

A Pair Of Glenn Curtiss Motorcycle & M-M Motorcycle Postcards,
ca. 1915. *Courtesy of the Doug Leikala Collection.* $25-$100 each

A Trio Of Cycling Postcards, two from MM, one from American Motor Company, ca. 1915. *Courtesy of the Doug Leikala Collection.* $20-$35 each

Miss Clara Wagner On Wagner Motorcycle Postcard, ca. 1920s. *Courtesy of the Doug Leikala Collection.* $20-$75

A Pair Of Postcards With Bikes And Riders, ca. 1920s. *Courtesy of the Doug Leikala Collection.* $20-$30 each

Firestone Ink Blotters, ca. 1920s. *Courtesy of the Doug Leikala Collection.* $25-$125 each

A Pair Of Indian Postcards, ca. 1910. *Courtesy of the Doug Leikala Collection.* $20-$30 each

A 1910 Harley Postcard, Harley Airplane Postcard. *Courtesy of the Doug Leikala Collection.* $20-$40 each

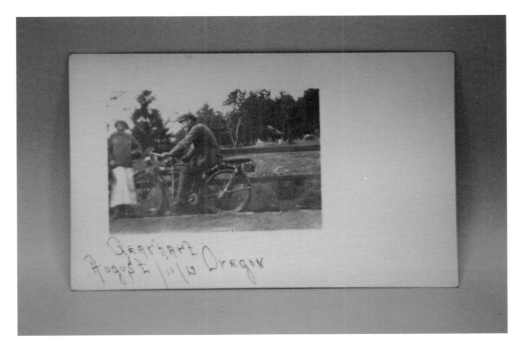

Indian Motorcycle Postcard, marked Gearhart Oregon, 1913. *Courtesy of the Dunbar Moonlighter Collection.* $10-$40

A 1913 Indian Motorcycle Club Picnic Postcard, three riders enjoying lunch on the beach, one wearing Indian sweater. *Courtesy of the Dunbar Moonlighter Collection.* $10-$40

A 1913 Indian Motorcycle Postcard, rider on bike coming out of garden, watch those petunias! *Courtesy of the Dunbar Moonlighter Collection.* $10-$40

Pair Postcards — "Ready For A Race," ca. 1910. Bike With Dog In Sidecar, ca. 1915. *Courtesy of the Doug Leikala Collection.* $10-$30 each

A 1933 German TT Races Postcard. *Courtesy of the Dunbar Moonlighter Collection.* $10-$40

A 1913 Indian Motorcycle Postcard, motorcyclists in pub or clubhouse, with AMC banner, Indian shirts, copy of Motorcycle Magazine. *Courtesy of the Dunbar Moonlighter Collection.* $10-$40

A Pair Of Upper Sandusky, Ohio, Postcards with riding black couple, ca. 1915. *Courtesy of the Doug Leikala Collection.* $30-$75 each

A Pair Of Postcards — Harley 1930s factory, 1940 Harley. *Courtesy of the Doug Leikala Collection.* $20-$30 each

A Pair Of Harley Davidson Postcards, ca. 1920s & 1950s. *Courtesy of
the Doug Leikala Collection.* $20-$50, $40-$100

A Pair Of Postcards — "Birthday Greetings" & "Happy Birthday," ca.
1915. *Courtesy of the Doug Leikala Collection.* $10-$30 each

"We're on the road to..." Postcard, ca. 1950s. "I can go some when I
get started...," ca. 1915. *Courtesy of the Doug Leikala Collection.*
$20-$30 each

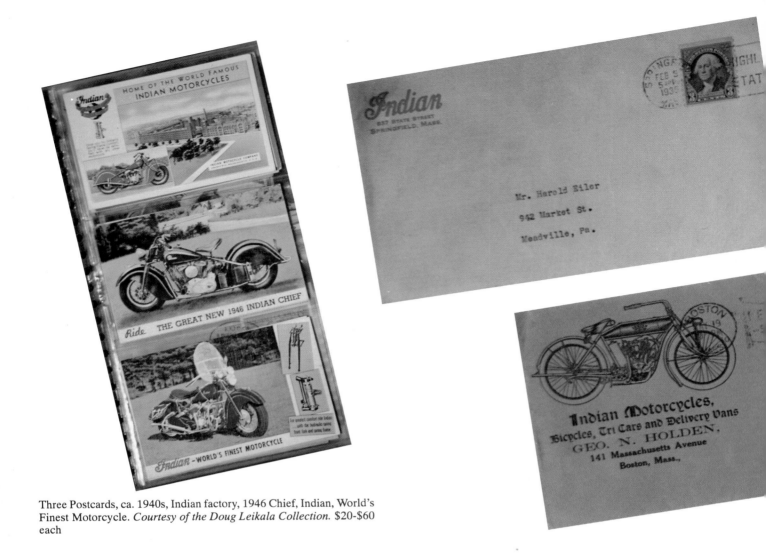

Three Postcards, ca. 1940s, Indian factory, 1946 Chief, Indian, World's
Finest Motorcycle. *Courtesy of the Doug Leikala Collection.* $20-$60
each

Indian Motocycle Stamped Envelopes, ca.1913-1920s. *Courtesy of the
Doug Leikala Collection.* $50-$125

The 1925 Series Of Airplane/Motorcycle Stamps. *Courtesy of the Doug
Leikala Collection.* $10-$35 each

A Trio Of Indian, Harley & Stamped Envelopes, ca. 1930s. *Courtesy of the Doug Leikala Collection.* $20-$50 each.

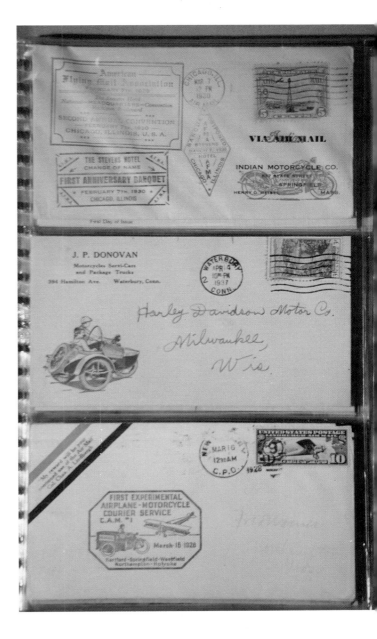

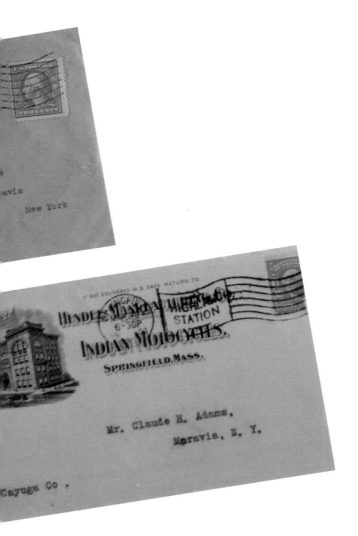

Variety Of Envelopes With Advertising Logos, ca. 1930s-1950s. *Courtesy of the Doug Leikala Collection.* $15-$35 each

Park Region Motorcycle Club Photo, 1913. *Courtesy of the Doug Leikala Collection.* $100-$250

A Photo Of Two Motorcyclists On Harleys, ca.1918. *Courtesy of the Doug Leikala Collection.* $50-$150

Aug. 11, 1946 McKeesport Gypsy Tour Photo. *Courtesy of the Doug Leikala Collection.* $50-$150

Gypsy Tour Held By Middle Atlantic Motorcycle Dealers Association Photo, Pike Tree Park, Aug. 23, 1938. *Courtesy of the Doug Leikala Collection.* $75-$150

Great Lakes Motorcycle Club Tour Photo, 1939. *Courtesy of the Doug Leikala Collection.* $50-$100

Annual Gypsy Tour Motorcyclists Photo, McKeesport motorcyclists, 1947. *Courtesy of the Doug Leikala Collection.* $75-$200

August 7, 1949 Annual Gypsy Tour Photo, McKeesport cyclist to West View Park, Pennsylvania. *Courtesy of the Doug Leikala Collection.* $100-$200

A Photo Of Early Motorcycle Shop, ca. 1915. *Courtesy of the Doug Leikala Collection.* $50-$150

A Photo Of Motorcyclists In Front Of "Gloves" Building, ca. 1915. *Courtesy of the Doug Leikala Collection.* $50-$150

Neracar Framed Photo, Cannonball Baker, ca. 1920s. Cannonball Baker was a mediocre racer but a determined long distance rider who crossed the United States several times on different bikes. *Courtesy of the Dunbar Moonlighter Collection.* $100-$300

NEW YORK TO LOS ANGELES ON 45 GALLONS OF GAS

"Cannon Ball" Baker Cuts Into Standard Oil Profits.

Erwin G. "Cannon Ball" Baker, famous racing driver and winner of many speed and endurance runs, has lowered another transcontinental record. This time, Baker has punctured the economy record by taking a 13½ cubic inch Neracar from New York to Los Angeles, 3,364.2 miles on 45 gallons of gas and 44.75 pints of oil at a fuel cost of $15.79.

Baker's running time of 174 hours and one minute (or 7 days, 6 hours and one minute) gives him an average of 19.3 miles per hour and 74.76 miles per gallon. The Neracar, made at Syracuse, N. Y., is the smallest vehicle to make the trip under its own power. It was Baker's sixtieth transcontinental trip.

PENNANTS & BANNERS

FAM National Convention Pennant, July 21-25, Sacramento, ca. 1915. Sponsored by Goodrich Tires. Any early FAM items are scarce and valuble, particularly such a dynamic and fragile pennant. *Courtesy of the Doug Leikala Collection.* $500-$1200

CHAPTER VII

Rah, rah, go team, wave that banner. Well, these pennants aren't of that 1920s raccoon coat, college type. Unlike fobs and pins, pennants and banners could be tacked on walls, large and bright for all to see. Much more difficult to find than paper or fobs, given their smaller production and greater fragility, pennants and banners offer a great backdrop to a collection. Doug Leikala's 1915 FAM pennant leads the way. It combines early FAM with Goodrich, one of the early makers of motorcycle tires that still exists (Goodyear is the other), with great graphics, color and condition; it is a winner in any circle. Note that the AMA gave out Club Safety Award Banners, ostensibly for the least number of accidents, broken bones, or missed red lights.

Embroidered FAM Felt Patch, ca. 1915. *Courtesy of the Doug Leikala Collection.* $100-$300

Embroidered Excelsior Felt Patch, ca. 1915. *Courtesy of the Doug Leikala Collection.* $100-$300

Langhorne, Pennsylvania Motorcycle Races Pennant, ca. 1940s. *Courtesy of the Doug Leikala Collection.* $100-$200

Midway Park, Chautauqua Lake Gypsy Tour 1953 Pennant. *Courtesy of the Doug Leikala Collection.* $75-$175

Harley Davidson Pennant, ca. 1920s. *Courtesy of the Doug Leikala Collection.* $300-$500

Harmstead & Holding Indian Motorcycle Pennant, ca. 1930s. Urbana, Ohio, with Crazy Indian logo. *Courtesy of the Doug Leikala Collection.* $400-$900.

Harley Davidson Pennant With Twin Cylinder Cycle, ca. 1914-1919. *Courtesy of the Doug Leikala Collection.* $300-$700

Harley Davidson Pennant, yellow with red lettering, ca. 1950s. *Courtesy of the Doug Leikala Collection.* $75-$150

Harley Davidson Pennant, yellow with navy lettering, ca. 1950s. *Courtesy of the Doug Leikala Collection.* $75-$150

Harley Davidson Pennant in navy and orange, ca. 1940s. *Courtesy of the Doug Leikala Collection.* $100-$250

Harley Davidson Pennant, navy & white, ca. 1950s. *Courtesy of the Doug Leikala Collection.* $75-$150

Harley Davidson Pennant, navy & white, ca. 1920s. *Courtesy of the Doug Leikala Collection.* $200-$400

Harley Davidson Pennant, grey with red lettering, ca. 1920s. *Courtesy of the Doug Leikala Collection.* $200-$500

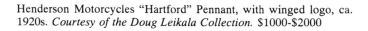

Henderson Motorcycles "Hartford" Pennant, with winged logo, ca. 1920s. *Courtesy of the Doug Leikala Collection.* $1000-$2000

The Indian Motocycle Crazy Indian Felt Logo, ca.1920s. *Courtesy of the Doug Leikala Collection.* $600-$1200.

Firestone Cycle Tires Pennant, purple & gold with FAM insignia, ca. 1915. *Courtesy of the Doug Leikala Collection.* $300-$700

The 1939 23rd Annual Gypsy Tour Pennant, Laconia New Hampshire.
Courtesy of Bob "Sprocket" Eckardt. $100-$350

The 1946 26th Annual Gypsy Tour Pennant, Laconia New Hampshire.
Courtesy of Bob "Sprocket" Eckardt. $100-$350

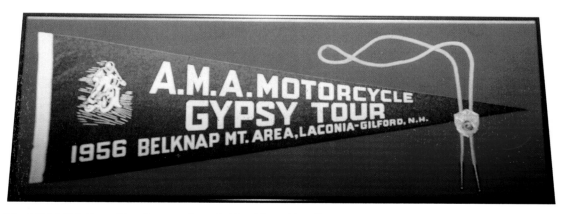

The 1956 AMA Gypsy Tour Pennant, Laconia New Hampshire. *Courtesy of Bob "Sprocket" Eckardt.* $100-$350

The 50th Golden Anniversary Harley Davidson Paper Banner, maroon
& yellow, 1954. *Courtesy of the Doug Leikala Collection.* $200-$500

The 1973 Indian Rally Day Banner. *Courtesy of the Doug Leikala
Collection.* $25-$75

The 1954 AMA Safety Award Banner. *Courtesy of the Doug Leikala Collection.* $75-$125

The 1957 AMA Safety Award Banner. *Courtesy of the Doug Leikala Collection.* $75-$125

The 1959 AMA Safety Award. *Courtesy of the Doug Leikala Collection.* $75-$125

RACING POSTERS & TROPHIES

"Indian Wins" Factory Poster, dated 1940, detailing Indian's victorious run at the 100 Mile National Championship Time Trial Race, Chattanooga, Tennessee, 34" W x 22" H. *Courtesy of the Dunbar Moonlight Kid Collection.* $200-$600

The history of motorcycle racing could constitute a book of its own. Almost as soon as motors were ignited, racing began. One of the first track races was held at Agricultural Park in Los Angeles in 1902, and was won by an Orient motorcycle. Racing soon expanded across the country, formally run by the FAM; informal races popping up wherever two or more motorcycles were together. Later on hill-climbs and time trial races became fashionable. Trophies from the earliest races, particularly those with a racer and a motorcycle brand on them, are the most desirable. National race trophies are also more valuable than local trophies. Racing posters were sometimes run with the same graphics for several years in a row. If they have nice graphics and color, they are both collectible and still reasonable. Even more collectible are early and unusual posters heralding a specific race.

AMA Trophy, loving cup, plastic base, metal top, incised AMA, 12" H. *Courtesy of the Dunbar Moonlighter Collection.* $50-$150.

Motorcycle Storecard, Medota, Illinois, June 22, ca. 1936-39, 14" W x 22" H. *Courtesy of the Dunbar Moonlighter Collection.* $100-$250

Set Of AMA Endurance Run Directional Cards, ca. 1950s. *Courtesy of the Doug Leikala Collection.* $50-$125 Set

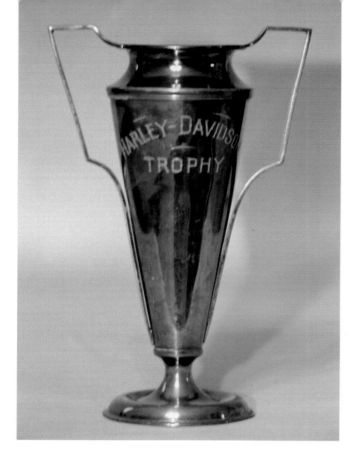

Harley Davidson Trophy, ca. 1915. *Courtesy of the Doug Leikala Collection.* $200-$450

First Place Trophy, TT Races, Gasden Alabama, won by Herman Dahlke, October 23, 1938. *Courtesy of Bob "Sprocket" Eckardt.* $100-$300

First Place Trophy, TT Race, Macon Georgia, won by Herman Dahlke, October 1, 1938. *Courtesy of Bob "Sprocket" Eckardt.* $100-$300

Singing Hills TT Races Storecard, featuring dueling Harleys, ca. 1936-39. 14" W x 11" H. *Courtesy of the Dunbar Moonlighter Collection.* $100-$300

First Place Brass Trophy, Georgia Crackers TT Race, 1939. *Courtesy of Bob "Sprocket" Eckardt.* $150-$350

Oshkosh Motorcycle Races Storecards, June 4, racing Harley cycles, ca.1936-39. 14" W x 22" H. *Courtesy of the Dunbar Moonlighter Collection.* $50-$250

Motorcycle Races Today Arrowcard, ca. 1950s. *Courtesy of the Doug Leikala Collection.* $ 50-$125

Official Ribbons Watkins Glen, 1958-1961. *Courtesy of the Doug Leikala Collection.* $25-$50 each

Cleveland Indian Motorcycle Club Trophy, ca. 1940s. *Courtesy of the Doug Leikala Collection.* $100-$250

Thrills — Spills Goodyear Storecard, Roanoke Fairgrounds, orange, black & white, with racer, 9" W x 15" H. *Courtesy of the Dunbar Moonlighter Collection.* $50-$100

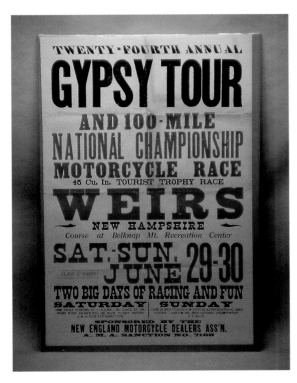

Gyspy Tour Championships Poster, dated 1940, Belnap Mountain, Weirs, New Hampshire, 25" W x 38" H. *Courtesy of the Dunbar Moonlight Kid Collection.* $100-$500

A 1956 AMA Trophy, large plastic dome with racer, marked inside base "1st Place Garden Spot 1956," 10 1/2" H x 7 1/2" Dia. *Courtesy of the Dunbar Moonlighter Collection.* $75-$200

AMA Trophy, small dome with racer, ca. 1950s, AMA logo. *Courtesy of the Dunbar Moonlighter Collection.* $50-$150

Motorcycle Races Storecard, Dover, New Hampshire, ca. 1950s. Sept. 25, 14" W x 22" H. *Courtesy of the Dunbar Moonlighter Collection.* $100-$250

Thompson Speedway Motorcycle Races Poster, July 4, 1940, 22" W x 28" H. *Courtesy of the Dunbar Moonlight Kid Collection.* $200-$600

AMA Trophy, ca. 1950s, wood triangle with racer, AMA logo, 9" H. *Courtesy of the Dunbar Moonlighter Collection.* $50-$150

AMA Trophy With Four Eagles & Racer, plastic base, metal top, ca. 1950s, 15" H. *Courtesy of the Dunbar Moonlighter Collection.* $50-$200

A 1953 Motorcycle Races Storecard, Granite State Park, Dover, New Hampshire, April 19, AMA sanctioned, 14" W x 22" H. *Courtesy of the Dunbar Moonlighter Collection.* $100-$250

Motorcycle Races Poster, Sturbridge Fairgrounds, Sunday, June 16, 1940, 14" W x 21" H. *Courtesy of the Dunbar Moonlight Kid Collection.* $50-$200

A 1957 AMA Trophy, gold triangle with racer, wood base, metal top, 14" H. *Courtesy of the Dunbar Moonlighter Collection.* $50-$200

Class "C" Motorcycle Hill Climb Poster, Sunday, Sept. 18th, Comet Motorcycle Club, ca. 1950s. *Courtesy of the Doug Leikala Collection.* $50-$200

Motorcycle Hill Climb Sample Advance Poster, ca. 1930s, 14" W x 20" H. *Courtesy of the Dunbar Moonlight Kid Collection.* $100-$300

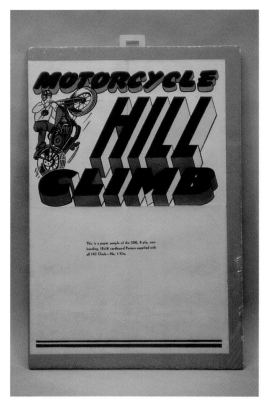

Motorcycle Hill Climb Sample Advance Poster, ca. 1930s, 12" W x 18" H. *Courtesy of the Dunbar Moonlight Kid Collection.* $100-$300

Motorcycle Gypsy Tour Poster, 23rd Annual AMA Meet, Aug. 11, 1940, 14" W x 22" H. *Courtesy of the Dunbar Moonlight Kid Collection.* $50-$200

1956 AMA Wood Trophy With Clock, 1st Place AMA Activity Contest, 9" x 9". *Courtesy of the Dunbar Moonlighter Collection.* $75-$250

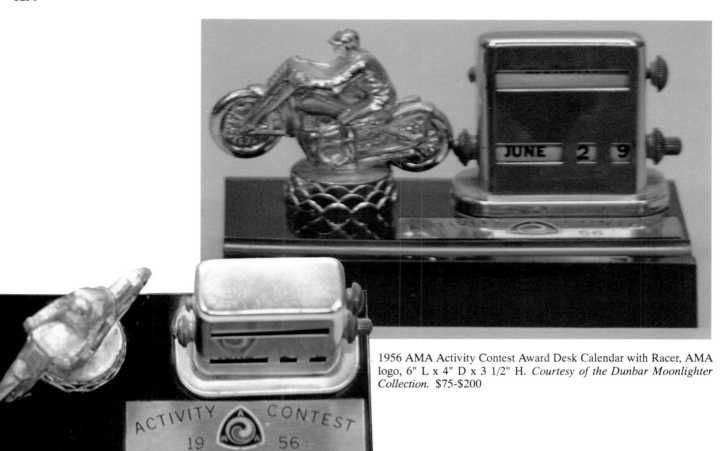

1956 AMA Activity Contest Award Desk Calendar with Racer, AMA logo, 6" L x 4" D x 3 1/2" H. *Courtesy of the Dunbar Moonlighter Collection.* $75-$200

Motorcycle Races Poster, July 4, 1940, officially AMA sponsored, Milwaukee, Wisconsin, 21" W x 27" H. *Courtesy of the Dunbar Moonlight Kid Collection.* $50-$300

AMA Trophy With Silver Racer, ca. 1950s, AMA logo, 7 1/2" H. *Courtesy of the Dunbar Moonlighter Collection.* $50-$150

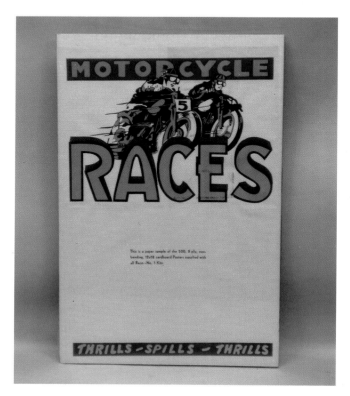

"Thrills — Spills — Thrills" Motorcycle Races Sample Advance Poster, ca. 1930s, 12" W x 18" H. *Courtesy of the Dunbar Moonlight Kid Collection.* $100-$300

Motorcycle Racing Storecard, Aug. 5, ca. 1960s 14" W x 22" H. *Courtesy of the Dunbar Moonlighter Collection.* $50-$150

National Championship Races Storecard, Hartford, May 30, ca. 1930s, 12" W x 30" H. *Courtesy of the Dunbar Moonlight Kid Collection.* $50-$200

AMA Motorcycle Scramble Races Storecard, July 19, ca. 1950s, 14" W x 22" H. *Courtesy of the Dunbar Moonlighter Collection.* $50-$200

AMA Trophy With Gold Climbing Racer, ca. 1930s. *Courtesy of the Doug Leikala Collection.* $100-$250

Harley Davidson Savings Bank Trophy, ca. 1940s. *Courtesy of the Doug Leikala Collection.* $100-$300

Pair Motorcycle Races Storecards, Pecatonica Illinois, AMA sanctioned, ca. 1950s. *Courtesy of the Doug Leikala Collection.* $50-$150

Motorcycle Club Advertising Sample Kit, original folder & mailing envelope, including samples of storecards, tickets, official & entrant tags, miniatures of race signs, sales letter, order form, reply card, 15 Pieces. *Courtesy of the Dunbar Moonlight Kid Collection.* $50-$250

New England Championship Motorcycle Races Poster, Old Orchard Beach, Maine, July 27-28, 1940, 14" W x 22" H. *Courtesy of the Dunbar Moonlight Kid Collection.* $50-$300

AMA Races Storecard, Aug. 8 & Aug. 28, ca. 1950s, AMA logo, 14" W x 22" H. *Courtesy of the Dunbar Moonlighter Collection.* $50-$150

Motorcycle Races Arrow, ca. 1950s. *Courtesy of the Doug Leikala Collection.* $20-$50

Motorcycle Races Storecard, ca. 1950s, Dover, New Hampshire, April 22, TT races, AMA sanctioned, 14" W x 22" H. *Courtesy of the Dunbar Moonlighter Collection.* $75-$200

Championship Races Storecard, Aug. 7, Allentown Fair, 19" W x 10" H. *Courtesy of the Dunbar Moonlighter Collection.* $50-$150

"Thrills" Motorcycle Races Sample Advance Poster, ca. 1930s, 14" W x 22" H. *Courtesy of the Dunbar Moonlight Kid Collection.* $100-$300

Motorcycle Hillclimb Stoneland, McKeesport, Pennsylvania. June 4, AMA sanctioned, ca. 1950s. *Courtesy of the Doug Leikala Collection.* $50-$200

Motorcycle Hillclimb Storecard, McKeesport, Pennsylvania, June 4, AMA sanctioned, ca. 1940s. *Courtesy of the Doug Leikala Collection.* $50-$200.

HATS & CLOTHES

Harley Davidson/Wrangler Cycle Wear Banner, ca. 1970s. A marriage made in hog heaven. *Courtesy of the Doug Leikala Collection.* $100-$450

Biker clothes are led by legendary leather, a scrape resistant (we hope), durable material that also happens to look cool most of the time. Black leather jackets with zippers, brown bomber jackets, suede with fringes — all are suitable gear for hitting the road. Add matching leather pants or jeans, a Harley "captain's" hat or a helmet, some kick-butt boots, shades, and an independent attitude, and it won't matter if you're Dennis Hopper or a head cleaner, you're a biker. Indian and Harley made and sold clothes to their riding customers. Sometimes dealers could make more money on clothes and accessories than on bikes and repairs. After all, we humans have an innate need to look the part, and to quote Dolly Parton in the movie *Steel Magnolias,* "accessories do make the outfit." Also included in this section is a great Motor Maids jersey. The Motor Maids organized in 1940 and joined the AMA in 1941. They still exist today as one of the premier women's cycling organizations.

Harley Davidson Riding Sweater With Embossed Insignia, ca.1930s. *Courtesy of the Doug Leikala Collection.* $200-$600

Harley Davidson Sweater With Insignia, ca. 1930s. *Courtesy of the Doug Leikala Collection.* $200-$600

Harley Davidson Sweater, maroon with insignia, ca.1930s. *Courtesy of the Doug Leikala Collection.* $200-$600

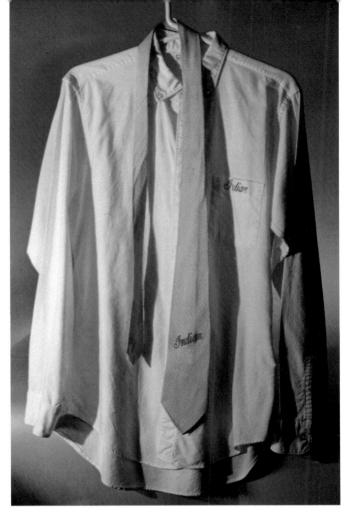

Indian Employee Collared Shirt & Tie, both embroidered Indian, ca. 1920s. *Courtesy of the Doug Leikala Collection.* $200-$600

Indian Sweater With Pocket Insignia, maroon & grey, ca. 1930s-1940s. *Courtesy of the Doug Leikala Collection.* $200-$600

Harley Davidson V-Neck Sweater With Insignia, ca. 1930s. *Courtesy of the Doug Leikala Collection.* $200-$600

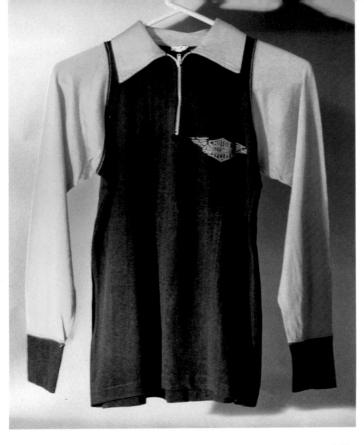

Harley Davidson Black & Orange Riding Jersey With Logo, ca. 1940s. *Courtesy of the Doug Leikala Collection.* $200-$600

Indian Red & White Riding Jersey, ca. 1940s. *Courtesy of the Doug Leikala Collection.* $200-$600

Harley Davidson Black & White Jersey, ca. 1940s. *Courtesy of the Doug Leikala Collection.* $200-$600

Judy - Motor Maid Of America Riding Sweater, front view, ca. 1940s-1950s. *Courtesy of the Doug Leikala Collection*. $200-$650.

Judy - Motor Maid Of America Riding Sweater, rear, ca. 1940s-1950s. *Courtesy of the Doug Leikala Collection*. $200-$650.

Harley Davidson Red Silk Jersey, ca. 1950s. *Courtesy of the Doug Leikala Collection*. $200-$600

Motor Maids From Kansas Denim Blouse, ca. 1950s. *Courtesy of Bob "Sprocket" Eckardt*. $50-$250

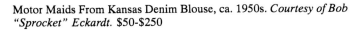

Indian Riding Jersey, white & red, ca. 1940s. *Courtesy of the Doug Leikala Collection.* $200-$600

Harley Davidson Racing Jersey, black & white, ca. 1940s. *Courtesy of the Doug Leikala Collection.* $200-$600

Indian Motorcycle Riding Jersey, ca. 1950s. *Courtesy of the Doug Leikala Collection.* $100-$300

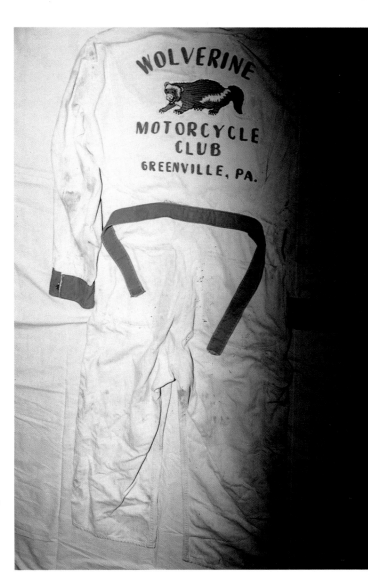

Embroidered ECR Motorcycle Club Inc. Jersey, Fostoria, Ohio, ca. 1950s. *Courtesy of the Doug Leikala Collection.* $100-$300

Wolverine Motorcycle Club Pit Jumpsuit, ca. 1950s. *Courtesy of Bob "Sprocket" Eckardt.* $100-$400

Sioux Valley Cycle Club Denim Racing Jacket, ca. 1950s. *Courtesy of Bob "Sprocket" Eckardt.* $50-$250

Indian T-Shirt, ca. 1950s. *Courtesy of the Doug Leikala Collection.* $75-$300

Harley Davidson Black Leather Riding Jacket And Pants, with 50th Anniversary patch, Sturgis, South Dakota, ca. 1960s. *Courtesy of Barry and Arline MacNeil.* $100-$350

Harley Davidson Sales Jacket, Cleveland, Ohio, black & white, ca. 1950s. *Courtesy of the Doug Leikala Collection.* $100-$300

Vandergrift Motorcycle Club Racing Jacket, ca. 1960s. *Courtesy of Bob "Sprocket" Eckardt.* $50-$150

Harley-Davidson Brown Leather Riding Jacket And Pants, ca. 1950s.
Courtesy of Barry and Arline MacNeil. $200-$500

AMF-Harley-Davidson Cloth Riding Jacket, ca. 1970s. *Courtesy of Barry and Arline MacNeil.* $50-$100

Pair Black Leather Riding Gloves, with studs and reflectors, ca. 1940s. *Courtesy of Barry and Arline MacNeil.* $100-$300

Motorcycle Races Tie, Sturgis, South Dakota, ca. 1940s. *Courtesy of Bob "Sprocket" Eckardt.* $50-$200

Pair Black Leather Riding Gloves, ca. 1930s, horsehair lining. *Courtesy of Barry and Arline MacNeil.* $100-$300

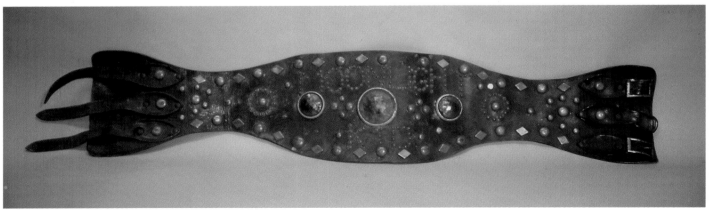

Moonlight Kid Leather Studded Kidney Belt, ca. 1930s, with reflectors. *Courtesy of the Dunbar Moonlight Kid Collection and the Cris Sommer/Pat Simmons Collection.* $200-$450

Harley Davidson NOS Motorcycle Goggles, ca. 1930-40, original box. *Courtesy Of the Ken Kalustian Collection.* $100-$300.

Harley Davidson Leather Kidney Belt, ca. 1930s-40s. *Courtesy Of the Ken Kalustian Collection.* $100-$400

Pair Harley Davidson Boots, insignias on soles, ca.1930s-40s. *Courtesy of the Doug Leikala Collection.* $100-$250.

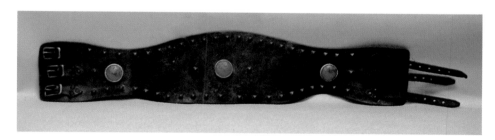

Brown Leather Studded Kidney Belt, ca. 1950s. *Courtesy of Barry and Arline MacNeil.* $100-$200

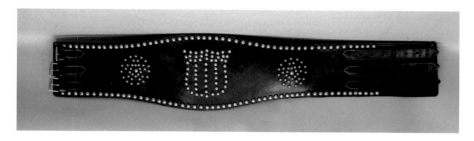

Brown Leather Studded Kidney Belt, ca. 1940s, with studs and reflectors. *Courtesy of Barry and Arline MacNeil.* $100-$200

Harley Davidson Leather Tooled Belt, ca. 1950s. *Courtesy of the Doug Leikala Collection.* $50-$100

Pair Motorcycle Leather Helmets, Frank Leahy & Western, ca. 1940s. *Courtesy of the Doug Leikala Collection.* $50-$150 each

Harley Davidson Riding Brimmed Cap With Brocade And Insignia, ca. 1950s. *Courtesy of the Dunbar Moonlighter Collection.* $75-$175.

(5) Harley Davidson Cop Style Riding Hats In Various Colors, ca. 1940s-1950s. *Courtesy of the Doug Leikala Collection.* $75-$125 each

Harley Cap With Insignia & Brocade, plain rider's hat, Harley winter's fur cap, ca. 1950s. *Courtesy of the Doug Leikala Collection.* $75-$150 each

(3) Harley Davidson Riding Cop Style Riding Hats In Various Colors, ca. 1940s-1950s. *Courtesy of the Doug Leikala Collection.* $75-$125 each

"You'll Look Still Better In a New Harley Davidson Cap" Mirror, ca. 1949. *Courtesy of the Doug Leikala Collection.* $300-$750.

(5) Harley Davidson Cop Style Riding Hats In Various Colors, ca. 1940s-1950s, *Courtesy of the Doug Leikala Collection.* $75-$125 each

(4) Harley Davidson Corduroy Brimmed Caps, ca. 1950s. *Courtesy of the Doug Leikala Collection.* $75-$125 each

Three Brimmed Rider's Caps With Winged Wheel Logo, ca. 1950s. *Courtesy of the Doug Leikala Collection.* $50-$150 each

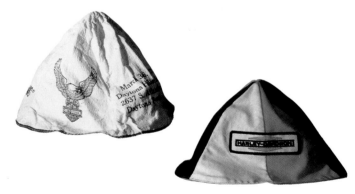

A Pair Of Harley Hats, white & purple, orange, black & white, ca. 1960s & 1980s. *Courtesy of the Doug Leikala Collection.* $20-$40

Harley Rider's Jerseys, Boots And Helmet. *Courtesy of the Doug Leikala Collection.*

TOOLS, ACCESSORIES & MISCELLANEOUS COLLECTIBLES

Pair Harley Davidson Motorcycle Sales Tags, ca. 1915-1920s. *Courtesy of the Doug Leikala Collection.* $100-$300 each

Like most mechanical or motorized inventions in their infancy, motorcycles were not quite as dependable as they are today. Stopping was usually achieved by using the Fred Flintstone method (feet, feet, feet) until brakes were developed. Sometimes fuel would stop flowing easily from engine to transmission, clutches would slip, chains would let go, and the cycle would stop. It helped to own a set of biker tools and it helped to have something to carry them in with all of a rider's other stuff.

Hence, saddle bags were introduced to the motorcycle and are still used today. Decorated saddle bags are very collectible, as are tools marked with makers' names, such as Harley or Indian. License plates, especially early ones, are also very desirable. And there are collectibles that really don't fall into any category, but they're neat. My favorite is the collapsible Indian spectator seat, suitable for watching your favorite racer or setting up and enjoying a canvas sit for a spell.

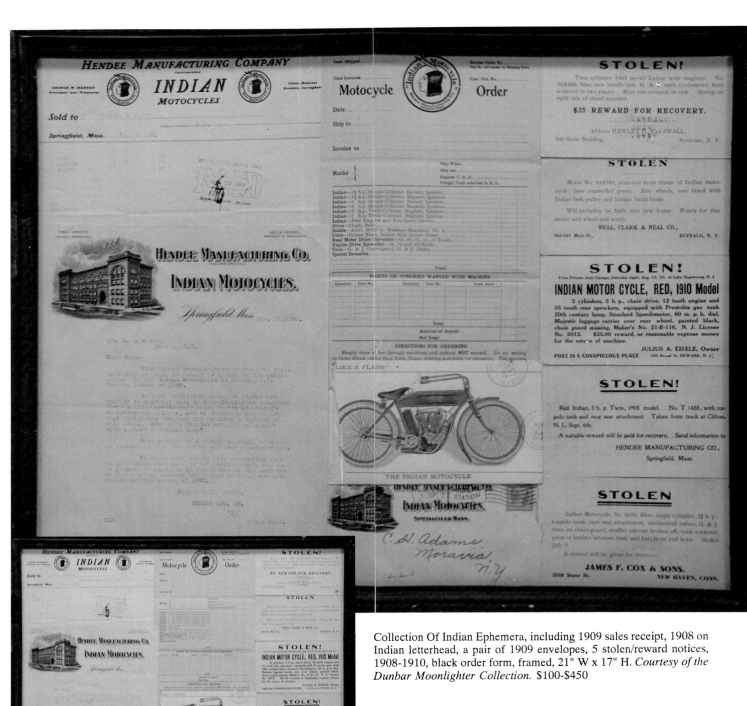

Collection Of Indian Ephemera, including 1909 sales receipt, 1908 on Indian letterhead, a pair of 1909 envelopes, 5 stolen/reward notices, 1908-1910, black order form, framed, 21" W x 17" H. *Courtesy of the Dunbar Moonlighter Collection.* $100-$450

FAM Motorcycle Reward Decals, ca. early 1900s. *Courtesy of the Doug Leikala Collection.* $100-$300 set

A Pair Of 1910 And 1911 Indian Sales Receipts, both to Claude Adams, 1910 for 5 HP Twin, 1911 for 7 HP Twin. *Courtesy of the Dunbar Moonlighter Collection.* $25-$50 each

Group Of Motorcycle Business Cards, 1920s-1960s. *Courtesy of the Doug Leikala Collection.* $50-$200

Indian Decals, ca. 1930s-40s. *Courtesy of the Dunbar Moonlighter Collection.* $25-$75 each

Motor Maids Of America Decal, Motor Maids National Convention Ribbon, 1955. *Courtesy of Bob "Sprocket" Eckardt.* $25-$75 each

Full Speed Needle Motorcycle Book. ca. 1950s. *Courtesy of the Doug Leikala Collection.* $25-$75

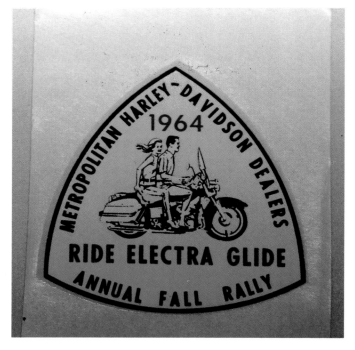

A 1964 Harley-Davidson Fall Rally. *Courtesy of Bob "Sprocket" Eckardt.* $25-$50

Indian Backpack, Indian Portable Canvas Seat, ca. 1930s. Great for spectating or sitting around the campfire. *Courtesy of the Doug Leikala Collection.* $350-$750.

Indian Taillight In Box, both light and box. *Courtesy of the Dunbar Moonlighter Collection.* $50-$100

Motorcycle Saddlebag, with fringe and studs. ca. 1940s. *Courtesy of Barry & Arline MacNeil.* $150-$300

Motorcycle Saddlebags, ca. 1940s. *Courtesy of Barry & Arline MacNeil.* $150-$300

Motorcycle Saddlebag, ca. 1940s. *Courtesy of Bob "Sprocket" Eckardt.* $200-$300

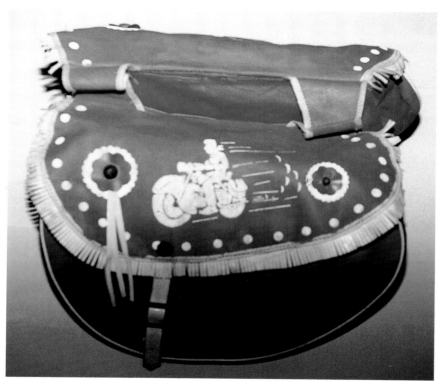

A Pair Of 1948 Massachusetts Motorcycle License Plates, #5703,
maroon with white. *Courtesy of the Dunbar Moonlighter Collection.*
$25-$75

A Pair Of 1951 Massachusetts Motorcycle License Plates, #4787.
Courtesy of the Dunbar Moonlighter Collection. $25-$75

A Pair Of 1953 Massachusetts Motorcycle License Plates, #2662.
Courtesy of the Dunbar Moonlighter Collection. $25-$75

A Pair Of 1955 Massachusetts Motorcycle License Plates, #4121.
Courtesy of the Dunbar Moonlighter Collection. $25-$75

Harley Chain Breaker And James Wrench, ca. 1920. *Courtesy of the Doug Leikala Collection.* Chain Breaker, $50-$125, Wrench, $20-$25

Pair Harley Chain Breaker Tools, ca. 1930s. Both are embossed with the Harley Davidson name. *Courtesy of the Dunbar Moonlighter Collection.* $50-$100 each

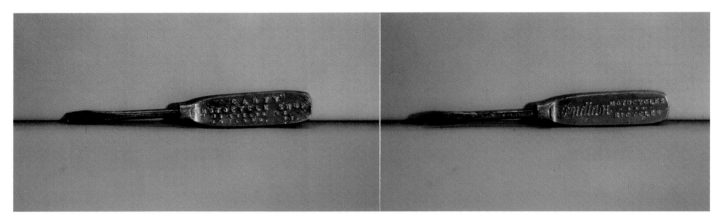

Indian Motocycles & Bicycles Souvenir Screwdriver, ca. 1920s, Canty Motocycle Shop, Rutland, Vermont, 3 1/2" L. *Courtesy of the Dunbar Moonlighter Collection.* $25-$75

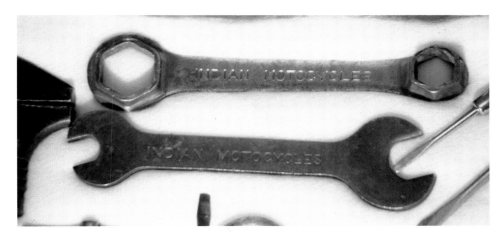

Indian Motocycles Wrenches, ca. 1920s. *Courtesy of the Doug Leikala Collection.* $20-$35

Harley Davidson Screwdrivers, c. 1950s, $20-$30 each. Reading Standard Motorcycle Screwdrivers, c. 1915-1920, $75-$175. Indian Motorcycle Wrench, c. 1920s, $50-100. *Courtesy of the Doug Leikala Collection.*

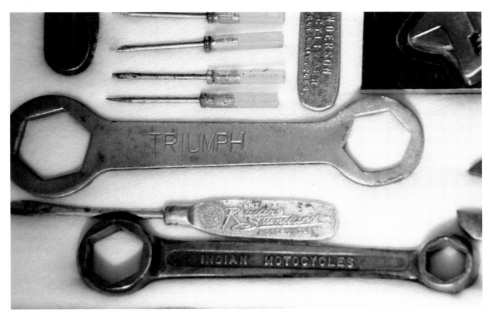

Henderson Motorcycle Screwdriver, c. 1920s, $75-150. Harley Davidson Wrench, c. 1950s, $50-150. *Courtesy of the Doug Leikala Collection.*

A 1971 Chrome & Hot Leather Lobbycard, 14" W x 11" H. *Courtesy of the Dunbar Moonlighter Collection.* $50-$150

Harley Davidson Wine Cooler, 1983-84, no longer produced; Liquor Decanter, Black Hills Motorcycle Rally. *Courtesy of Ken Kalustian.* $10-$15; $50-$150

TOYS

Hubley Indian "Say It With Flowers" Cast Iron Delivery Van, ca.1930s, removable driver, 10 1/2" Long. Hubley produced this cycle, part of Indian's Servi-Car series, in five different sizes, all of which are extremely collectible. This model is the largest and hardest to find. There is even a motorized version, which is seldom seen. *Courtesy of the Dunbar Moonlight Kid Collection.* $10000-$20000

Just like automobiles, the excitement of motorcycling caught the imagination of a generation catching the first major wave of the industrial era. If kids weren't old enough to hop on the back of motorized bikes, at least they could wind up or push a miniature model down the driveway. The 1920s-'30s were the heyday for American tin windup toys. Mostly produced by Marx, they featured colorful litho bodies which — for a few pennies — whirred, stopped, fell over, and righted themselves (unless of course an errant tree or table leg was in the way). Their heavyweight counterparts, made of cast iron, were built primarily by the Hubley Manufacturing Company of Lancaster, Pennsylvania. Amazingly foresighted, Hubley signed an exclusive agreement with Indian and Harley-Davidson to produce motorcycle toys based on real life models. The result was long lived and profitable for all parties. Harley even offered some of the toys in the *Enthusiasts*. Years later, these pieces are the most collectible of motorcycling toys, with the biggest and best examples reaching into thousands of dollars, an investment way beyond child's play.

Hubley Indian Cast Iron Traffic Car, ca. 1930s, 12" Long, removable driver. Real life traffic cars expedited equipment and lights to organize heavy traffic areas. *Courtesy of the Dunbar Moonlight Kid Collection.* $2200-$7500

Hubley Indian US Air Mail Cast Iron Delivery Van, ca. 1930s, removable driver, 8 1/2" Long. What could say more about a company's dependability than being entrusted with the daily mail service? And I bet this carrier was smiling. *Courtesy of the Dunbar Moonlight Kid Collection.* $1000-$3500

Hubley Indian Cast Iron Crash Car, ca. 1930s, removable driver, complete with accessories cans, axes, hosereel, 11 1/2" Long. Crash cars, the forerunners to EMT's, raced to the scene of automotive accidents to aid injured motorists and disabled vehicles. *Courtesy of the Dunbar Moonlight Kid Collection.* $2200-$7500

Hubley Indian Cast Iron Armored Police Cycle With Sidecar, ca.1930s, removable driver & passenger, 8 1/2" Long. Police used armored cars to shield them from the submachine gun fire of America's most wanted of the 1920s-'30s; Capone, Dillinger, et al. *Courtesy of the Dunbar Moonlight Kid Collection.* $1000-$3000

Hubley Indian Cast Iron Police Cycle With Sidecar, ca. 1930s, removable driver & passenger, 8 1/2" Long. *Courtesy of the Dunbar Moonlighter Collection.* $1000-$2000

Hubley Indian Cast Iron Cycle W/Sidecar, ca. 1930s, removable civilian driver & passenger, 8 1/2" Long. Hubley made both Harley & Indian motorycles with sidecars with civilian figures as well as police figures. However, the civilian figures are much rarer. *Courtesy of the Dunbar Moonlighter Collection.* $3000-$7500

Hubley Indian 4 Cylinder Police Solo Cast Iron Motorcycle, ca. 1930s, removable driver, 9 1/4" Long. *Courtesy of the Dunbar Moonlight Kid Collection.* $600-$2200

Hubley Harley Davidson Cast Iron Police Motorcycle With Sidecar, with driver & passenger, ca. 1930s, 9" Long. Sales of cycles to local police departments helped to keep Harley-Davidson in business during the Depression. *Courtesy of the Dunbar Moonlight Kid Collection.* $800-$2200

Hubley Indian Cast Iron Traffic Car, ca.1930s, swivelhead driver, 9" Long. *Courtesy of the Dunbar Moonlight Kid Collection.* $500-$2200

Hubley Harley Davidson Cast Iron Police Solo Motorcycle, removable driver, ca. 1930s, 8 1/2" Long. *Courtesy of the Dunbar Moonlight Kid Collection.* $500-$1800

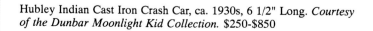

Hubley Indian Cast Iron Crash Car, ca. 1930s, 6 1/2" Long. *Courtesy of the Dunbar Moonlight Kid Collection.* $250-$850

Hubley Harley Davidson Parcel Post Cast Iron Delivery Motorcycle, ca. 1930s, with removable driver, 9 1/2" Long. Harley and Indian competed for the service delivery bike market as much as for the police and civilian business. *Courtesy of the Dunbar Moonlight Kid Collection.* $1200-$3000

Hubley Harley Davidson Cop CI Solo Motorcycle, orange, 8 1/2" Long, original driver, handlebars, decals, ca. 1930s. *Courtesy of the Doug Leikala Collection.* $800-$2200.

Hubley Cast Iron Speed Racing Cycle, ca. 1930s, #7, 6" Long. Kids could vicariously feel the thrill of competing on the short track. *Courtesy of the Dunbar Moonlight Kid Collection.* $400-$2000

Hubley Harley Davidson Cast Iron Hillclimber Motorcycle, ca.1930s, #45, 6 1/2" Long. Hillclimb races captured the imagination (and kidneys) of the cycling world during the Depression. Clubs just had to find a relatively clear hill to, well, climb. This model of toy bike was one of Hubley's most popular and is fairly easy to find, usually in well raced condition. Really super examples are scarce. *Courtesy of the Dunbar Moonlight Kid Collection.* $300-$1300.

Hubley Harley Davidson CI Solo Civilian Motorcycle, ca. 1930s, 6" Long. This toy was made in several colors, including orange, red, dark blue, and olive drab green. Somehow, the cap always stays on. *Courtesy of the Dunbar Moonlight Kid Collection.* $300-$1800

Hubley Harley Davidson Jr. Cast Iron Motorcycle, ca. 1930s, 5 1/2" Long. $250-$750. Hubley Harley Davidson Military Solo Cast Iron Motorcycle, ca. 1930s, 5 1/2" Long. *Courtesy of the Dunbar Moonlight Kid Collection.* $250-$650

Hubley Harley Davidson Cast Iron Civilian Motorcycle & Sidecar, ca.1930s, with civilian removable passenger, 5 1/2" Long. *Courtesy of the Doug Leikala Collection.* $600-$1800

Hubley Harley Davidson Cast Iron Police Motorcycle & Sidecar, ca. 1930s, 5 1/2" Long. *Courtesy of the Dunbar Moonlight Kid Collection.* $500-$1500

Hubley Harley Davidson Solo Cast Iron Police Motorcycle With Swivelhead Driver, ca. 1930s, 7" Long. *Courtesy of the Dunbar Moonlight Kid Collection.* $500-$1100

Hubley Cast Iron Windup Cast Iron Motorcycle & Sidecar With Driver, clockwork mechanism. Very few vehicles, let alone motorcycles, were motorized — probably because the expense in shipping was greater than the extra cost gained. *Courtesy of James J. Julia Auctions, Inc.* $5000-$15000

Hubley Cast Iron Police Solo Motorcycle, ca. 1930s, 8 1/2" Long, blue color variation. *Courtesy of the Dunbar Moonlight Kid Collection.* $400-$1200

Hubley Popeye Cast Iron Solo Patrol Cycle, ca.1930s, removable Popeye figure, 9" Long. In the late 1930s, Hubley decided to make Popeye a biker, joining Clark Gable, Van Johnson and other riding stars. Produced in 3 versions, for some reason, the cycle never caught on and is tough to find today, especially since it is also treasured by comic character collectors. Beware of reproductions. *Courtesy of the Dunbar Moonlight Kid Collection.* $2500-$7500

Hubley Popeye Cast Iron Spinach Wagon, ca. 1930s, 5 1/2" Long. This toy is much easier to acquire and much less expensive. *Courtesy of the Dunbar Moonlight Kid Collection.* $400-$1100

Vindex Excelsior-Henderson 4 Cylinder Cast Iron PDQ Delivery Van, circa 1930, removable driver, 9" Long. Vindex made sewing machines, but chose to enter the cast iron toy market just before the stock market crash. Irwin Schwinn bought Excelsior-Henderson from the Henderson brothers and ran a small division, carried by the profits of his bicycle line. *Courtesy of the Dunbar Moonlight Kid Collection.* $2000-$6500

Hubley Cast Iron Police Solo Motorcycle, ca. 1930s, 8 1/2" Long. Although Hubley owned the exclusive patent rights to manufacture replicas of both Indian and Harley, the company also made a number of generic bikes, which have Hubley decals only. *Courtesy of the Dunbar Moonlight Kid Collection.* $400-$1200

Vindex Excelsior-Henderson 4 Cylinder Cast Iron Cycle With Sidecar, circa 1930, removable driver, 8 1/2" Long. All Vindex cycles are difficult to find, because Vindex went out of the toy business in 1931. *Courtesy of the Dunbar Moonlight Kid Collection.* $2000-$4500

Vindex Excelsior-Henderson 4 Cylinder Cast Iron Solo Cycle, circa 1930, removable driver, 8 1/2" Long. *Courtesy of the Dunbar Moonlight Kid Collection.* $800-$2500

Champion Cast Iron Cop Cycle & Sidecar, ca. 1930s, 5" Long. *Courtesy of the Dunbar Moonlight Kid Collection.* $500-$1400

Champion Cast Iron Cop Cycle & Sidecar, ca. 1930s, 6 1/4" Long.
Courtesy of the Dunbar Moonlight Kid Collection. $900-$1800

Hubley Indian "Say It With Flowers" Cast Iron Delivery Van, ca. 1930s, 3 1/2" Long, the smallest size. *Courtesy of the Doug Leikala Collection.* $600-$2000

Globe Cast Iron Police Motorcycle, ca.1930s, 8 1/2" Long. Globe was a small company that made two different police cycles, a solo and a cycle with a sidecar. *Courtesy of the Dunbar Moonlight Kid Collection.* $1000-$2000

Vindex Excelsior-Henderson 4 Cylinder Cast Iron Solo Cycle, circa 1930, removable driver, 8 1/2" Long, red variation. *Courtesy of the Dunbar Moonlight Kid Collection.* $800-$2500

Champion Cast Iron Cop Cycle, ca. 1930s, nickel wheels, 5" Long. $100-$450. Champion Cast Iron Cop Cycle, ca. 1930s, 7 1/4" Long. $200-$750. Champion Cast Iron Cop Cycle, ca. 1930s, rubber tires & wheels, 5" Long, $75-$300. *All Courtesy of the Dunbar Moonlight Kid Collection.*

Champion Cast Iron Cop Cycle, 1930s, 7 1/4" Long, unusual green color variation. This bike also came in red, which is seldom seen. *Courtesy of the Dunbar Moonlight Kid Collection.* $300-$950

Pair Hubley Cast Iron Police Patrol Motorcycles, ca. 1930s, 6 1/2" Long. *Courtesy of the Dunbar Moonlight Kid Collection.* $150-$500 each

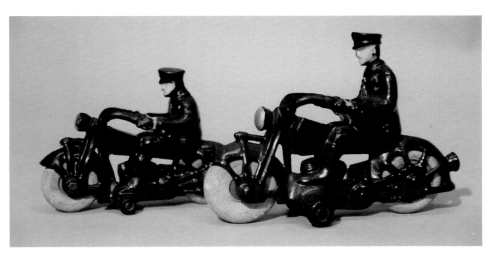

AC Williams Cast Iron Cop Cycle, 1930s, 4 7/8" Long. $150-$450; AC Williams Cast Iron Cop Cycle, 1930s, 6 1/2" Long. *Courtesy of the Dunbar Moonlight Kid Collection.* $200-$650

AC Williams Cast Iron Cop Cycle, ca. 1930s, 6 1/2" Long. Unusual orange color variation. *Courtesy of the Dunbar Moonlight Kid Collection.* $350-$850

Hubley Tandem PDH Cast Iron Motorcycle, ca.1930s, 4" Long. $50-$125. Kilgore Special Delivery Cast Iron Motorcycle, ca. 1930s, 4 1/2" Long. *Courtesy of the Dunbar Moonlight Kid Collection.* $100-$450

Hubley Police Solo Cop Cast Iron Motorcycle, ca. 1930s, electric headlight, 6 1/2" Long. *Courtesy of the Dunbar Moonlight Kid Collection.* $150-$650 each.

Hubley Cast Iron Cop Motorcycle And Sidecar, ca. 1930s, removable passenger, 4" Long. *Courtesy of the Dunbar Moonlight Kid Collection.* $50-$175. Hubley Cast Iron Solo Cop Cycle, ca.1930s, 4" Long. *Courtesy of the Dunbar Moonlight Kid Collection.* $25-$150

Hubley Cast Iron Cop Solo Cycle, ca.1930s, 3" Long. $25-$100. Hubley Cast Iron Cop Solo Cycle, ca.1930s, 4 1/4" Long. $50-$125. Hubley Traffic Cast Iron Car, ca. 1930s, 4 3/4" Long. *Courtesy of the Dunbar Moonlight Kid Collection.* $75-$350.

Mickey & Minnie Tin Windup Motorcycle, made by Tipp Co., Germany, ca. early 1930s. 9" Long. Although this book is about American collectibles and the toy is German, what can be more red, white, and blue than Disney's favorite mice? Check out the great rat faces and autobahn smiles. *Courtesy of James J. Julia Auctions, Inc.* $15000-$30000

Hoge Tin Windup Traffic Cycle Car Delivery Van, ca. 1930s, 10" Long.
Courtesy of James J. Julia Auctions, Inc. $1000-$3500

Marx Speed Boy Motorcycle Delivery With Electric Light & Box, ca. 1930s. Marx made the first version of this bike with a battery powered headlight. Subsequent runs omitted the light and changed the litho color scheme of the bike and rider. *Courtesy of the Ken Kalustian Collection.* $750-$1600 boxed

Marx Speed Boy Tin Windup Motorcycle Delivery With Box, ca. 1930s,
9 1/2" Long. *Courtesy of the Dunbar Moonlighter II Collection.* $600-
$1500 boxed

Marx Tin Windup Siren Police Cycle, ca. 1930s, with box, $300-$800.
Vinyl Dare Devil Cycler Motorcycle Rider, ca. 1950s, Thomas. $20-
$35. *Courtesy of the Ken Kalustian Collection.*

Pair Marx Speed Boy Tin Windup Motorcycle Delivery Cycles, ca.
1930s, 9 1/2" Long. *Courtesy of the Doug Leikala Collection*. $300-
$850 each unboxed.

Marx Tin Windup Siren Cop Cycle, ca. 1930s, 8" Long. Wind up this bike, the siren blows, the cop takes off, and pulls you over ... just like in real life. *Courtesy of the Dunbar Moonlighter Collection.* $100-$650.

Pair Marx Police Tin Windup Cycles, ca. 1930s, both 8" Long. *Courtesy of the Doug Leikala Collection.* $100-$550 each.

Marx Tin Windup Tipover Cop Cycle, ca. 1930s, 8" Long. Wind up this bike, he rolls, tips over, and rights himself ... just like in real life. *Courtesy of the Dunbar Moonlighter Collection.* $100-$450

A Pair Of Model Cycles, *Courtesy of the Doug Leikala Collection.*
$50-$150 each.

A Pair Of Duo-Glide Motorcycles, Courtesy of Doug Leikala Collection. $50-$150 each.

Plastic Space Cycle, ca. 1950s. *Courtesy of the Doug Leikala Collection.* $100-$350

A Group Of Plastic Motorcycles: featuring a witch on a motorcycle, a large cop motorcycle, and smaller cop motorcycles, ca. 1940s-1960s. $35-$150 each. *Courtesy of the Doug Leikala Collection.* The witch on a motorcycle is also a popular Halloween collectible and sells between $100-$250.

Group Plastic Cycles, ca. 1960s. *Courtesy of the Doug Leikala Collection.* $20-$100

Group Of Plastic Motorcycles With Sidecars, ca.1940s-1960s. *Courtesy of the Doug Leikala Collection.* $35-150 each

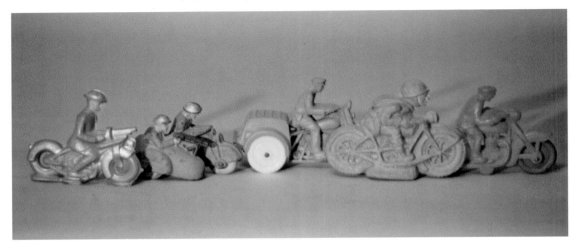

Group Of Five Motorcycles, plastic, etc., 1940s-1960s. *Courtesy of the Doug Leikala Collection.*

Motor Cycle Jr. Police (wink wink, nudge nudge) Gun, Badge and Handcuff Set, boxed, ca. 1950s. *Courtesy of the Doug Leikala Collection.* $100-$350

Speed Kop Plastic Whistle On Original Card, 1950s, 4" Long, unused condition. *Courtesy of the Dunbar Moonlighter Collection.* $10-$75

Pair Of Slushcast Motorcycle & Bicycle, ca. 1920s. *Courtesy of the Dunbar Moonlighter Collection.* $20-$55 each

SOURCES

MOTORCYCLING COLLECTIBLES INFORMATION AND SOURCES

Motorcycling Organizations:
Antique Motorcycling Club Of America
Dick & Wanda Winger, Membership Chairpersons
PO Box 333
Sweetster, IN 46987
Publication: The Antique Motorcycle

American Motorcyclist Association
33 Collegeview Rd.
Westerville, OH 43081-1484
Publication: American Motorcyclist

Publications:
American Rider
TL Enterprises
3601 Calle Tecate
Camarillo, CA 93012
(805) 389-0300

Hemmings Motor News
Box 390, Route 9 West
Bennington, VT 05201
802-442-3101

Mobilia
Eric Killorin, Tom Funk
RD2, Box 4365, Lemon Faire Rd.
Middlebury, VT 05753
802-545-2510

Motorcycle Shopper Magazine
Luis Hernandez
1353 Herndon Ave.
Deltona, FL 32725
407-860-1989

Walneck's Classic Cycle-Trader
Buzz Walneck
923 Janes Ave.
Woodridge, IL 60571
708-985-2097

Motorcycle Collectibles Dealer, Direct Sales & Auctions:
Dunbar's Gallery
Howard Dunbar, Leila Dunbar
76 Haven St.
Milford, Massachusetts 01757
508-634-8697 phone, 508-634-8698 fax